Hydroponic Tomatoes

for the Home Gardener

Dedication

I wish to dedicate this book to Carolyn Weeks who has a keen interest in hydroponics and believes that it is my duty to make this information available to everyone.

Also, for the support she has given in making my previous books successful through her efforts at Woodbridge Press in their promotion and sales.

So ... to Carolyn, who encouraged me and insisted that I write this book.

Thanks!

Hydroponic Tomatoes

for the

Home Gardener

By Howard M. Resh, Ph.D.

CRC Press
Taylor & Francis Group
Boca Raton   London   New York

CRC Press is an imprint of the
Taylor & Francis Group, an **informa** business

CRC Press
Taylor & Francis Group
6000 Broken Sound Parkway NW, Suite 300
Boca Raton, FL 33487-2742

International Standard Book Number-13: 978-0-931231-97-1 (Softcover)
Cover painting and text illustrations: Janice Blair

This book contains information obtained from authentic and highly regarded sources. Reprinted material is quoted with permission, and sources are indicated. A wide variety of references are listed. Reasonable efforts have been made to publish reliable data and information, but the author and the publisher cannot assume responsibility for the validity of all materials or for the consequences of their use.

Library of Congress Cataloging-in-Publication Data

Resh, Howard, M.
  Hydroponic tomatoes : For the home gardener.
     p. cm.
  Includes index.
  ISBN 0-931231-97-3 (alk. paper)
  1. Tomatoes. 2. Hydroponics. I. Title.
SB349.R35   1993, 2003
635'.64285—dc20   93-18128

Visit the Taylor & Francis Web site at
http://www.taylorandfrancis.com

and the CRC Press Web site at
http://www.crcpress.com

Contents

Appendix

List of Figures

Introduction

This book will assist you in the growing of tomatoes hydroponically for your own pleasure and use. Information is presented in easy-to-use terms and procedures so that even without previous experience in growing plants hydro-ponically you can understand and apply the methods described. It is not intended for the commercial grower. Commercial growers should seek a more advanced text such as *Hydroponic Food Production;* or, for more comprehensive home hydroponics guidance, *Hydroponic Home Food Gardens,* both by the author of this book and, like this one, published by Woodbridge Press.

The hydroponic growing of plants is *soilless culture*—the growing of plants without soil. A substitute "soil" medium is used, such as a mixture of peat, vermiculite (expanded mica), perlite (heated pumice) or styrofoam chips. These media are commonly sold at garden centers under the name of "potting mixes" for house-plants. However, soilless culture can also include the use of pure sand (of volcanic origin, not of sedimentary, seashell origin), coarse vermiculite or perlite alone or mixed, fir bark or other non-soil materials. Recently, rockwool (similar to fiberglass insulation (but having more rigid structure) has become very popular for commercial growers and can be easily adapted for use by the backyard gardener.

True hydroponic culture implies using water itself, with added "fertilizing" nutrients, to make up a solution to feed the plants. An example of water culture is the nutrient film technique (NFT), described on Page 53.

Any one of several such methods can be used successfully in the growing of tomatoes hydroponically. The objective of this book is to present various methods and cultural practices that you can readily use in growing tomatoes on a small scale at home.

The growing of your own tomatoes hydroponically will provide you with an enjoyable experience in nurturing healthy plants and will reward you with clean, attractive, and healthful food.

You will have the pleasure of growing your own tomatoes in an environment free of harmful pesticides. If you do encounter pests, you may control them by use of biological agents or safe natural pesticides, such as "Safer's Soap." These methods will be discussed in the section on cultural practices.

Using a balanced nutrient solution as described in this book, your plants will grow vigorously, allowing full assimilation of all essential elements needed to produce fruit rich in the vitamins and elements important to your own diet.

The hydroponic systems described in this book are illustrated in diagrams and photographs so that you can easily follow procedures to build them yourself.

If you wish, you may purchase small hydroponic units from any of a number of companies listed in the Appendix. These companies also sell prepared nutrient mixes. You can follow the package directions indicating the amount of nutrients to be added to a given volume of water to make up your nutrient solution. Or, you may make your own nutrient solution by following the nutrient formulations and purchasing the salts recommended in this book.

Whatever method, system or techniques you use, you may determine for yourself how involved you wish to get in the growing of your own tomatoes hydroponically and therefore how much time you would like to spend on this self-satisfying, productive hobby.

Good gardening; *let's get started!*

Chapter 1

Hydroponic Systems

for Superior Tomatoes

Pot Culture and Bag Culture

In the very simplest form of hydroponics, you can grow a tomato plant in a plastic or ceramic pot just as you would many houseplants. However, since tomato plants require more nutrients (they grow more vigorously than houseplants), you should use a larger pot than you would normally use for houseplants. The best size pot would be a deep, three-gallon or five-gallon nursery pot. These pots measure approximately 10 1/2 inches wide by 10 inches deep, and 12 inches by 11 inches for the three- and five-gallon pots, respectively.

A plastic saucer or tray should be placed underneath the pot to collect excess nutrient solution as it drains through the medium when you water the plant. This solution must be removed from the saucer after full drainage. Do not let the pot sit in the solution for any length of time or the medium growing your tomato plant will become saturated and cause suffocation of the plant roots due to lack of oxygen.

The Wick System

Another, and long-used, method of hydroponic pot culture is the wick system. This method uses one pot nested into another.

The lower pot acts as the nutrient reservoir, and a fibrous wick draws up the solution to nourish the plant growing in the upper pot.

The lower pot should have no holes. If you use a pot with holes as the reservoir, line it with black 6-mil-thick polyethylene sheet plastic, which you can buy at hydroponics shops or building supply stores.

Drill a hole in the center of the bottom of the upper growing pot and insert the wick snugly so that it will remain in place. Fray both ends of the wick to give it more surface area to improve solution uptake.

Fill the upper pot with a potting medium of peat, perlite and vermiculite as described below, or a porous gravel such as expanded clay. This is clay or shale fired at high temperatures to cause its expansion and formation into small rock particles from 1/4-inch to 1/2-inch in diameter. It has a characteristic reddish brown color. It is used in the landscaping industry and can be purchased at garden centers.

As you fill the pot, set the wick in place so that it is oriented vertically in the center of the potting medium. This will give uniform distribution of the nutrient solution in the potting medium as capillary action moves the solution vertically up the wick and laterally through the medium. Fill the pot to within one-half inch of the top, but keep the wick end two inches below the surface. The growing pot sits in the nutrient pot on a support keeping the base of the growing pot out of the solution below it (Fig. 1). This is to prevent excessive moisture in the potting medium which could lead to lack of oxygen to the plant roots. The support can be made from a piece of PVC plastic pipe. For example, for a five-gallon nursery pot having a 10 to 12-inch base, three pieces two inches long of 2-inch diameter pipe will work. Drill 1/4-inch holes spirally around the pipe from top to bottom so that the nutrient solution can pass through these pipe supports. The wick should be placed in the solution between the supports.

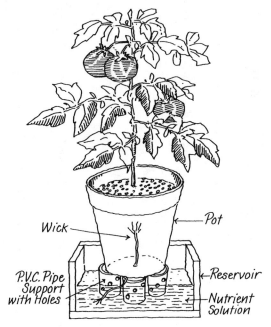

Figure 1. Wick system pot.

This wick system can be expanded into a patio planter by building a reservoir of wood, lined with 6 mil black polyethylene.

For a series of six 12-inch diameter nursery pots, for instance, construct a nutrient tank 30" by 44" by 4" high (Fig. 2). Make a cover of plywood painted with exterior white enamel paint or use a piece of white rigid or 6 mil polyethylene.

Cut holes in the cover at the locations of the pots. The diameter of the holes should be just large enough to allow the pots to go through. Stagger the position of the holes to get optimum use of space as shown in Figure 2. Place pot supports underneath each pot, using short pieces of PVC pipe as described above. A floating solution indicator can be placed in the nutrient tank to assist you in maintaining the solution level.

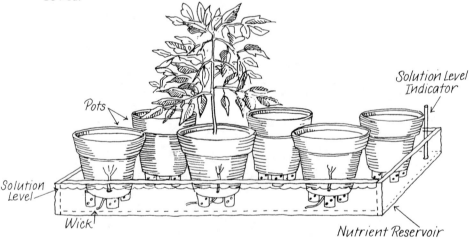

Figure 2. Several wick pots placed in a common reservoir.

Between biweekly changes of the solution, simply add water as required. Floating level indicators are available from garden centers and hydroponic supply stores (see Appendix).

The Self-Watering Planter

Another kind of pot hydroponic system available commercially from garden shops is the self-watering planter commonly used for houseplants. It can be used as a hydroponic planter with a peat-lite medium as described below or with expanded clay rock.

While these wick- and self-watering planters can grow tomatoes, I prefer to use a regular single pot as described above because it allows better aeration of the growing medium by adding solution from the top and allowing complete drainage. This moves oxygen through the root system and leaches away any excess salts which might accumulate in the root zone, as can easily occur with self-watering systems.

A Simple Plastic Bag

While I prefer a rigid plastic pot, which is stronger and more aesthetically pleasing, you can use a simple plastic bag. A plastic bag of the "Kitchen Catcher" (eight-gallon) type may be used to hold the growing medium for your tomato plant.

Fill the bag with medium and set it on a plastic tray to collect drainage water (Fig.3). Punch four to five holes about one-quarter inch in diameter, using a paper punch, at the bottom of the bag, around its base.

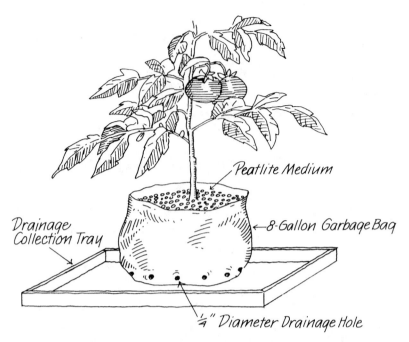

Figure 3. Eight-gallon plastic kitchen garbage bag with peatlite medium, growing a tomato plant.

An alternative is to make four slits of approximately one-half inch around the base of the bag at the bottom. Be sure that these holes or slits are immediately at the bottom of the bag on the sides so that complete drainage will occur when you water the plant.

Productive Watering

Watering the plant once a day should be sufficient unless it is in the summer sun during full growth. Then you should water it several times a day. If the plant wilts, it indicates that it is not getting sufficient moisture. This would occur especially during the hot afternoon period of the day. The plant should be watered with a complete nutrient solution, as presented later in this book under "nutrient solutions."

Water the plant enough to get at least a 20 percent drainage. That is, add enough nutrient solution so that 20 percent of the volume of water you apply will drain out. You can measure this by collecting the solution which accumulates in the pan underneath the plant grow-bag. This is important so that the solution applied leaches out any excess salt build-up from previous waterings.

Keep in mind that excess watering will cause root die-back and also wilting of the plant. But that would be caused by too-frequent watering and/or allowing the grow-bag or pot to stand in the drainage water.

The watering regimen should therefore be: Water only as frequently as is necessary, when the soilless medium dries somewhat. Then water enough to allow a 20 percent leaching; then remove any accumulated solution underneath the bag or pot in the tray or saucer.

If you wish to be more accurate in meeting the watering needs of your plants, you may use a water moisture sensor, available at any garden center. These sensors have a probe which you insert into the medium in the root zone to test the level of moisture present. The meter indicates whether there is excessive, sufficient or insufficient levels of moisture in the medium. That takes the guess work out of the watering.

Otherwise, you can use your finger to test the moisture content of the medium. With some experience you will soon learn how moist the medium is and when to water. It is important that when you test the moisture in the medium with your finger, it does not feel either very saturated or very dry. In general, overwatering not underwatering, is the most common error in caring for plants.

The "Soil" Growing Medium

Now, let's discuss in more detail the "soil" medium that you may use in growing tomatoes in pots or plastic bags. A potting mix purchased from your nearby garden center will be adequate for growing tomatoes because you will be adding the necessary nutrients with each watering with your nutrient solution.

Most potting mixes are composed of 60 percent to 75 percent sphagnum peat, 10 percent to 15 percent perlite, and the balance is made up of vermiculite. For a heavier mix, sand may be substituted for the vermiculite.

These commercially available mixes generally also have additives such as dolomite lime to adjust the pH (acidity/alkalinity—discussed later under "nutrient solutions"), a wetting agent to overcome the resistance of peat to absorbing water and various "time-release" fertilizers.

Time-release fertilizers are encapsulated within a water-permeable membrane which allows the fertilizers to escape slowly into the medium at low levels each time water is applied to the plant. Tomatoes, however, are very heavy feeders and do better if a complete nutrient solution is applied during each watering.

You may purchase these complete potting mixes in 3- or 4-cubic-foot bags, or you may make up your own.

If you wish to make your own, purchase a bag each of peat, perlite and vermiculite and mix them together at the above percentages on a sheet of polyethylene or a clean concrete floor.

Peat is very acidic, having a pH of 4.0 to 5.0. Its pH can be raised to the optimum level of 6.3 to 6.5 for tomatoes by adding some dolomite lime, which is available at a garden center. The lime is usually added at a rate of two pounds per cubic yard (27 cubic feet) or 1.2 ounces per cubic foot, the amount of medium contained in a space of 1 foot x 1 foot x 1 foot.

While mixing the lime and various medium components, add water. Add enough water, a little at a time, until a handful of medium, squeezed tightly, will retain its shape and start to crack slightly when you open your hand.

If excess moisture runs through your fingers as you squeeze the ball of medium, you have added too much water. In that case, adjust the moisture level by adding a little more dry peat. If the medium is too dry, the ball will fall apart as you slowly open your hand. Do this test several times while adding water to the medium until you get the correct moisture level.

Other media that may be used are: perlite, vermiculite or sawdust alone or a mixture of these. An equal mixture of all three will work, or approximately half perlite and half fine vermiculite, or an equal ratio of perlite and sawdust or vermiculite and sawdust.

To get better lateral movement of water along the surface of these particular media, it is advisable to add a one-half to one-inch layer of peat on the top of the medium in the growing container. In this way, the water will not simply channel through the very permeable coarse particles of the perlite, vermiculite or sawdust. With these media you need not add any dolomite limestone—except with sawdust alone, and only about half of what is needed with peat.

Your nutrient solution will provide all the essential elements, and, since these media are neutral in acidity, the pH will be adjusted through the application of the nutrient solution.

The Importance of Sunlight

Tomatoes require direct sunlight to be productive, so they should be placed in a sunny location on your patio or in your back yard. If you are growing them entirely inside, you must provide artificial lighting by the use of cool-white, high-output fluorescent lights. This may not be economically feasible, but if you desire fresh tomatoes during winter months or in areas having no direct sunlight, it is absolutely necessary to use supplementary artificial lighting.

Even with supplementary lighting, of course, your production will be somewhat less than it would be when growing in direct sunlight.

Grow-Bags

White polyethylene "grow-bags" will each grow three tomato plants (Fig.4). A grow-bag is a "layflat" plastic sleeve of 6 mil thickness, measuring one foot wide by three and one-half feet long. The sleeve is heat sealed on one end, filled with a perlite or sawdust medium, then heat sealed on the other end. When filled, the bags are 10 inches in diameter and three feet long.

The plants may be started in rockwool cubes (see section on rockwool culture) or in peat-lite medium (as described above). In the latter case, the medium is placed in four-inch-square band pots. These pots have a cross-strip on the bottom to contain the medium, yet allow the roots to grow out of the pot base.

Once the seedlings are about four weeks old, with roots emerging from the bottom of the pot, they can be transplanted.

Cut a hole in the top of the grow-bag where each transplant is to be located, then place the band pot or rockwool cube into the hole, sitting directly on top of the medium that is in the grow bag.

Irrigate the transplant from the top of the band pot or rockwool cube for several days to allow the roots to enter the grow bag, then irrigate at the base of the pot or cube, using a drip emitter to each plant (Fig.4).

If you use perlite as a medium in the bag, presoak the bag with nutrient solution for one day prior to transplanting.

Upon transplanting, cut a half-inch slit diagonally in the side of the bag facing the irrigation line. The slit should be about one to one and one-half inches up from the bottom. This maintains a reservoir of solution to that height in the bottom of the grow-bag. In this way, should your system fail at any time, there will be sufficient water in the bag to carry the plant a day or so

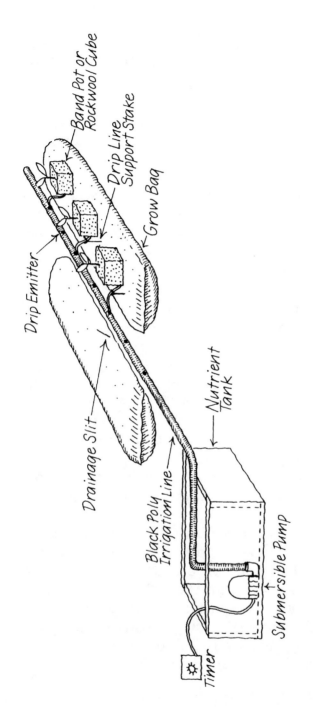

Band Pot or Rockwool Cube

Drip Line Support Stake

Grow Bag

Drip Emitter

Drainage Slit

Nutrient Tank

Black Poly Irrigation Line

Submersible Pump

Timer

Figure 4. Grow-bag open system—no return of solution.

24

depending upon how old the plant is and the temperature and light conditions at the time.

If sawdust is used in the bags, two slits should be made on the inside face of the bag to achieve sufficient drainage. The slits should be made at the bottom of the bag, not at one to one and one-half inches up from the bottom as with perlite (above). This is because sawdust has a much higher moisture retention than perlite, and slower drainage, so it will not dry out as fast as perlite and risks puddling (excessive moisture) should the slits not be made at the bottom of the bag.

An Irrigation System

This type of hydroponic system requires drip irrigation to apply the nutrient solution automatically, using a time-clock, nutrient reservoir and pump as shown in Figure 5.

For backyard gardening you should use a returnable or recycle system. In this type of closed system, the grow-bag(s) should be placed in a grow-channel made of plastic or wood, lined with polyethylene.

If the grow-bags are 10 inches in diameter, the grow-channel should be at least 12 inches wide on the inside to allow sufficient space for the nutrient solution to flow past the bags.

Tomato plants should be spaced about 14 to 16 inches apart within the rows.

To do this with the three-foot-long grow-bag, locate one plant in the center of the bag and the other two plants 14 to 16 inches to either side. That is, within two to three inches of each end of the bag. When positioning the bags end-to-end

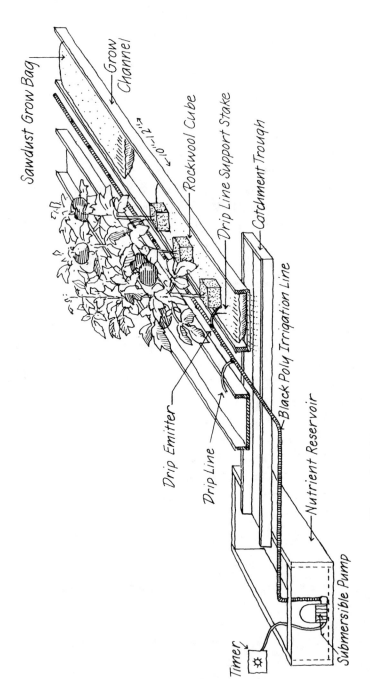

Sawdust Grow Bag

Grow Channel

Rockwool Cube

Drip Line Support Stake

Catchment Trough

Drip Emitter

Drip Line

Black Poly Irrigation Line

Nutrient Reservoir

Submersible Pump

Timer

←10"←12"→

Figure 5. Grow-bag closed system—return of solution.

26

in the grow-channel, place them 10 to 12 inches apart, thus giving a 14- to 16-inch spacing between the plants on the ends of the bags.

The sides of the grow-channel should be at least two to three inches high on the inside.

The channel can be easily constructed of 1-inch by 12-inch lumber with 1-inch by 4-inch sides glued and nailed at the joints (Fig.6). You should line this grow-channel with 6 mil black polyethylene or 20 mil vinyl. The black color will reduce algae growth where the nutrient solution is exposed to the light.

Paint the outside of the channel with a primer coat and two coats of white exterior enamel.

When placing the liner in the grow-channel, fold the corners in an envelope-like fashion. Staple the liner at the top edges of the channel. After placing and stapling the liner, cover the edges and top of the sides with ducting tape.

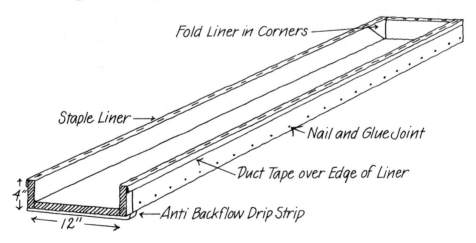

Figure 6. Grow-channel detail—sawdust bag system.

At the drain end secure the liner under the channel with staples and an aluminum anti-backflow strip.

Screw a 1/2-inch square aluminum tubing on the bottom about half an inch back from the edge as the anti-backflow strip to prevent the solution from flowing back under the channel. Seal it with a caulking compound such as silicone rubber sealant.

The channel should have a three percent slope towards the nutrient tank. A channel containing three three-foot grow-bags end-to-end, with 12 inches between them, would thus have a slope of about four inches.

If several rows are constructed, a catchment trough, perpendicular to the rows, will collect the nutrient solution runoff from the channels (Fig.7) and transport it to the nutrient tank. The catchment trough may be made from a length of four-inch plastic irrigation pipe, with a section cut out along the top to allow the nutrient solution runoff to enter the pipe. It must be covered with 6 mil white polyethylene to reflect heat and prevent algae growth.

When two rows of plants are grown, the plant spacing between rows should be about 16 inches. If more than two rows are grown, a 30-inch-wide aisle is needed between pairs of plant rows.

The Nutrient Tank

The size of the nutrient tank is a function of the number of plants you are growing. Allow at least two quarts of solution per plant per day, with adequate volume for at least five days. That is, if you have a dozen mature tomato plants growing, a nutrient tank of 12 x 5 x 2 = 120 quarts or 30 gallons in volume would be required.

A submersible pump of the type used in water fountains (for example, a "Little Giant") will be adequate if it is capable of pumping about 300 gallons per hour at a one-foot lift. It should have a pumping height capacity of 10 to 12 feet.

The pump is connected to a 1/2-inch diameter black polyethylene hose designed for drip irrigation systems. The black polyethylene hose carries the nutrient solution along the length of the beds and drip emitters are attached to it at the location of each plant (Fig.7).

A small one-quarter inch drip line runs from the emitter to the base of the plant-growing cube or pot on top of the growing bag. A small plastic stake supports the drip line as shown (Fig.7).

The pump is operated by a time-clock. An irrigation controller like those used in landscape irrigation systems serves this purpose well.

The time-clock or controller must have a 24-hour clock in series with a 60-minute clock so that a number of irrigation cycles may be set over the 24-hour period. Alternatively, a 24-hour timer and a 60-minute timer may be wired in series to get the same results at a lower cost. A hydroponics or irrigation system supplier can show you how to do this.

Irrigation cycles will have to be adjusted for the plant stage of growth and environmental conditions. For example, when plants are very young, not producing any fruit, three to four daily irrigations will be adequate. But during the mature fruiting stage, under full summer sunlight conditions, the frequency of irrigation cycles may be every hour during daylight and three to four times during the night. Simply, watch your plants; if they wilt from lack of moisture, increase the frequency of irrigation.

If you construct this grow-bag system in your backyard to grow tomatoes outside during the summer season, the nutrient tank can be partially buried to within two inches of the ground surface. That is just enough height to prevent any heavy rainstorm from flooding the tank. A cover is placed on top of the tank to keep out sunlight and rainfall. Any rain will enter the channels and flow into the nutrient tank, diluting the nutrient solution. When that occurs, you must adjust the nutrient solution proportionately according to the amount of rainfall that entered the tank (see section on nutrient solutions and the Appendix).

The channels will have to be supported with bricks, wood or a plastic PVC pipe frame in order to obtain the correct three percent slope towards the catchment trough or nutrient tank (Fig.7).

Some modifications to this system will be necessary when constructing it indoors or on a patio. The channels are raised above the level of the nutrient tank and high enough to get a three percent slope to the tank. This can be easily done with a PVC pipe frame of one and one-half inch or 2-inch diameter so that it is strong enough to support the weight of the channels and mature plants (Fig.8).

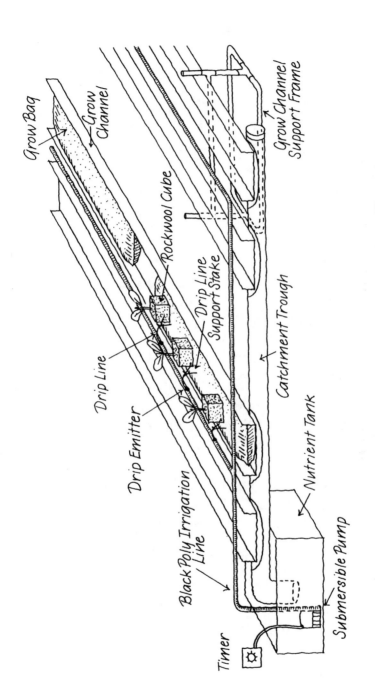

Grow Bag

Grow Channel

Rockwool Cube

Drip Line

Drip Line Support Stake

Drip Emitter

Black Poly Irrigation Line

Catchment Trough

Grow Channel Support Frame

Nutrient Tank

Timer

Submersible Pump

Figure 7. A series of grow-bags in a returnable system.

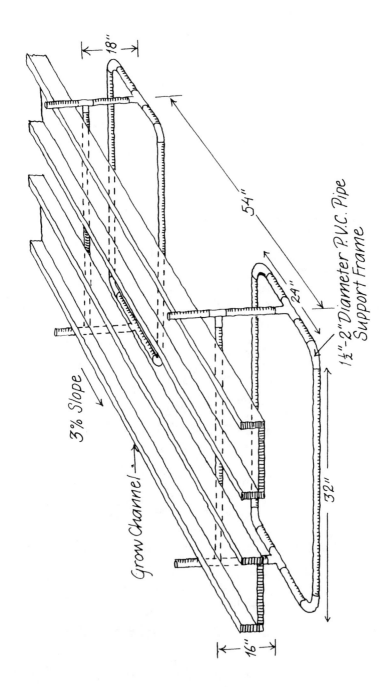

18″

54″

24″

1½″–2″ Diameter P.V.C. Pipe
Support Frame

32″

3% Slope

Grow Channel·

16″

Figure 8. Grow-bag system—grow-channel support frame
detail for a nutrient tank 12 inches high.

Place these support frames at four to five-foot centers. For a grow-bag system having grow-channels 11 feet long, containing three grow-bags, use three support frames: one, at one foot from the outlet end; the other in the center and the third, one foot from the inlet end.

For a three percent slope the height difference between outlet and inlet ends is four inches. Build three support frames, each two inches different in height. For a tank 12 inches high, the outlet, middle and inlet supports should be 16, 18 and 20 inches high, respectively.

This height at the outlet end will allow four-inch clearance between the ends of the channels and the top of the nutrient tank so that the catchment trough (pipe) will fit under the ends of the channels and still can be placed on top of the nutrient tank so that it may drain into the nutrient tank through its cover (Figs.7,8).

The width of the support frames is 32", allowing a 12-1/2" width for each growing channel and 6" between the channels. This will give 16"-18" between the rows of plants.

When constructing a larger system having a relatively large nutrient tank that is going to be located inside or on a patio, be careful to allow for the weight of the nutrient solution. It is like a water bed but more concentrated over a smaller space.

One U.S. gallon of water weighs 8.3 pounds, while one Imperial gallon weighs 10 pounds. Therefore, a 30-U.S.-gallon tank of nutrient solution will weigh: 8.3 x 30 = 250 lb. Such a 30-gallon tank is equivalent to four cubic feet which could measure 1 foot x 2 feet x 2 feet. Thus, the total 250 pounds is spread over an area of only four square feet, giving a pressure of 62.5 pounds per square foot. Most structures and patios can support such a weight, but you better check to be sure!

Small Indoor Units

Small indoor hydroponic growing units are available commercially, specifically designed for the home gardener. They may be used in the house, apartment or on a patio.

For tomatoes, it is preferable to place such a unit on the patio, or other location, where the plants can receive direct sunlight. During winter months the patio could be enclosed or a polyethylene frame placed over top of the hydroponic garden. If you wish to grow tomatoes during the cooler seasons of late fall to early spring you should put a small heater in the patio enclosure or under the polyethylene covering of the hydroponic unit.

Most small indoor units measure approximately one foot high by one foot wide by two feet long. They are double-chambered, having a lower nutrient reservoir with an upper tray holding the growing medium.

A common fish aquarium air pump outside of the unit pumps air through a small hose inserted into a slightly larger one inside the nutrient tank. The larger hose must be of sufficient diameter that there is an air space between it and the smaller hose from the air pump.

A pin holds the two hoses together where the one is inserted into the other by about one inch (Fig.9). This allows entrance of the nutrient solution into the union of the two hoses as air travels past it. The air pushes the nutrient solution up through the larger-diameter hose, which is perforated as it enters the upper tray and runs along the surface of the medium of the tray, irrigating the plants growing in it (Fig.10). A clear polyvinyl tubing of 3/8-inch outside diameter will serve as the irrigation line. Drill 1/16-inch diameter holes spaced two to three inches apart along the top surface of the tubing.

Generally, the best growing medium for the tray is a coarse vermiculite. Perlite can also be used, but its capillary action in carrying the nutrient solution laterally is not as strong as that of the vermiculite. Also, vermiculite retains more moisture than perlite. A coarse vermiculite and perlite mixture often does not have adequate capillary action. It can, however, be improved by placing a 1/4" to 1/2" layer of peat or fine vermiculite on top of the coarser medium.

The tray bottom supporting the medium is perforated so that excess solution will seep back into the nutrient reservoir below (Fig.10).

If you are making your own hydroponic unit, drill 1/4-inch holes at a 3" by 3" spacing in the bottom of the tray to get sufficient drainage. The depth of the tray is four to five inches. The nutrient reservoir is five to six inches in depth. These units are available complete from a number of hydroponic equipment and supply stores (see Appendix).

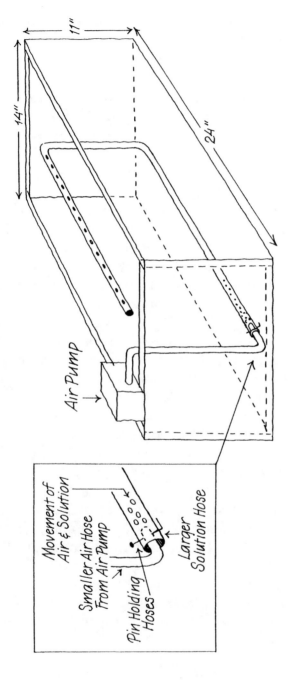

11"

14"

24"

Air Pump

Movement of
Air & Solution

Smaller Air Hose
From Air Pump

Pin Holding
Hoses

Larger
Solution Hose

Figure 9. Air pump moving solution up a tube, using the
tube-in-tube principle.

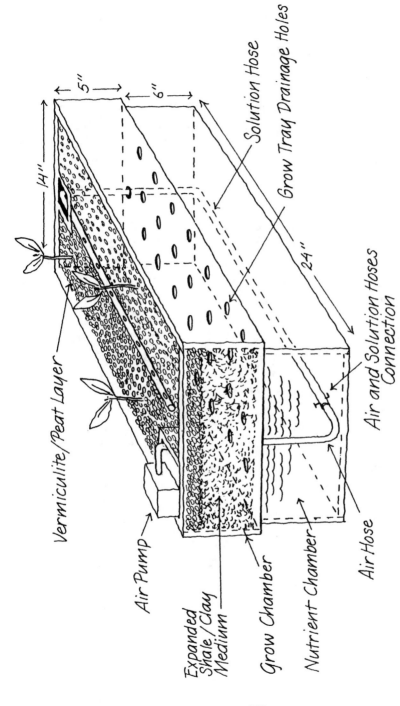

Figure 10. Small indoor double-chambered hydroponic unit.

37

Between crops, simply replace the medium entirely rather than trying to sterilize it. This will assure that no pests or diseases will be carried over from the previous crop. Sterilize the nutrient tank and surfaces of the hydroponic unit with a 10 percent bleach solution when changing your crop.

Plant Care

This type of hydroponic unit will support the growth of two to three tomato plants if you train them properly.

The stems of staking tomatoes need to be supported by stakes, a trellis, or support strings tied to a hook in the ceiling. Spread the stems of the upper portion of the plants so that they are at least 14 to 16 inches apart (Fig.11). If three plants are grown in the hydroponic unit, position one of them in the center, on one side of the unit, and the other two on the other side and at either end, some four inches in from the edge of the unit. Thus, the bases of the plants will only be eight inches apart. However, that is not a problem as there will be plenty of nutrients for the root systems, which will spread throughout the medium of the upper tray.

By separating the upper stems of the plants to a more optimum spacing, sufficient light enters the plants so that they will be productive. Methods of training and caring for your plants are discussed at greater length in the section on cultural practices.

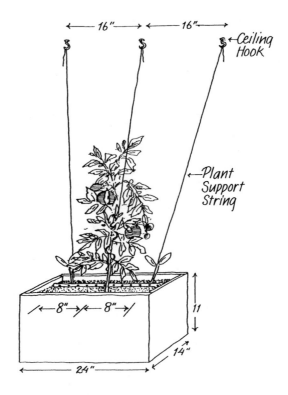

Figure 11. Support of tomato plants in a
small hydroponic unit.

The Nutrient Solution

The nutrient solution is prepared according to
the volume of the tank and formulation required
during the growing stages of the plant as
discussed in the later section on nutrient solutions. For best results you should change the
nutrient solution at least every other week. It
becomes unbalanced with time, as the plants
assimilate various elements at different rates.

Determine the volume of the tank, then refer
to Table 3 or Table 4 in the Appendix to calculate
the amounts of various fertilizer salts to add to
the water.

For example, work out the volume of the tank as follows:

Width = 12" Length = 24" Depth = 6"

Volume = Width x Length x Depth

= 12" x 24" x 6"

or: 1 foot x 2 feet x 1/2 feet = 1 cubic foot

1 cubic foot = 7.48 U.S. gallons

Therefore, total volume = 1 x 7.48 = 7.48 or about 7.5 U.S. gallons.

Now, follow Table 3 or Table 4 in the Appendix to determine the amounts of each nutrient salt required.

Using Various Hydroponic Cultures

The small tank-and-tray indoor hydroponic growing systems we have been discussing may be modified for various kinds of hydroponic culture, like: aggregate culture, ebb and flow rockwool culture and drip rockwool culture. The essential components of all such systems are: nutrient tank, pump, timer, growing bed and medium.

Aggregate Culture. In an aggregate culture system, in most cases, a light-weight, heat expanded, clay material is used. The particle sizes range from one-eighth inch to one-half inch in diameter. You can easily recognize this expanded clay by its reddish-brown color. It is commonly used as a medium in the self-watering pot culture systems as described earlier in the section on pot culture.

The nutrient solution tank may be of rigid A.B.S. plastic, fiberglass or wood construction lined with 20 mil vinyl (Fig. 12).

If the hydroponic unit dimensions are two feet

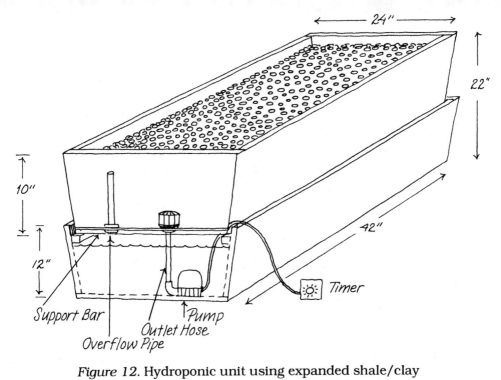

Figure 12. Hydroponic unit using expanded shale/clay medium, sub-irrigation or "ebb-and-flow" system.

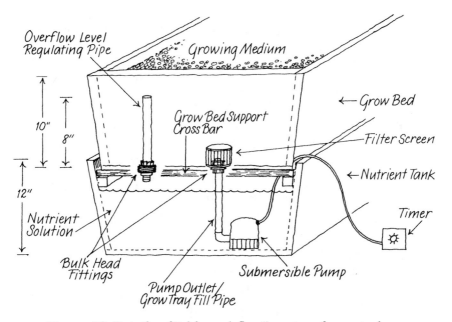

Figure 13. Details of "ebb-and-flow" system for an indoor hydroponic unit.

by three and one-half feet, the depth of the tank should be 10 to 12 inches. The volume of the tank would therefore be: V = Width x Length x Depth; V = 2 x 3.5 x 10/12 = 5.83 cubic feet (assuming it is 10 inches deep).

If one cubic foot = 7.48 U.S. gallons, the total volume in gallons is: 5.83 x 7.48 = 43.6 gallons. Note that you cannot fill the tank up to the brim, so you must allow at least one to two inches of extra height. That is, for 10 inches of usable depth in the tank, construct it 11 to 12 inches in height. Cross-support bars are placed in the upper part of the tank about one inch from the top (Fig.12). This supports the growing tray, in addition to holding the tank from bowing outward from the pressure of the solution when the tank is full. The growing tray then nests one inch inside the nutrient tank supported by the cross bars.

Ebb and Flow System. In an ebb and flow or "subirrigation" system the growing tray must be without holes. An overflow pipe regulates the correct solution level during a "feed cycle." This is a 3/4-inch diameter PVC plastic pipe secured into the base of the growing tray with a bulkhead fitting, a threaded fitting having rubber seals on either side (Fig.13).

The depth of the growing tray for tomatoes should be at least 10 inches so that there is a usable depth for medium of nine inches. The height of the overflow pipe should then be eight inches so that the solution does not moisten the surface of the medium. This will prevent algae growth and possible insect infestation by fungus gnats (see pests and disease section).

Complete drainage takes place by the solution draining back through the pump via the outlet pipe. Solution must not remain in the medium, as this will cause lack of oxygen (puddling) and subsequent root die-back.

The growing tray can be constructed of the same materials as the nutrient tank, but, the dimensions must be slightly smaller than the tank so that it will fit into the top of the tank. For example, if the tank dimensions are 24 by 42 inches, then the growing tray dimensions should be 23 by 41 inches.

A submersible pump of 80 gallons per hour capacity will provide adequate water flow for an indoor hydroponic unit with up to two feet by three to four feet feet of growing space. The submersible pump is attached to a PVC plastic pipe which is secured and sealed to the bottom of the growing tray by a bulk head fitting similar to the overflow pipe (Fig. 13). A filter screen is placed on top of the pump outlet pipe so that the medium directly above it will not enter the pump outlet.

The pump is operated by a 24-hour, 15-minute increment timer such as is used for turning lights on and off in your house. These are readily available from a building supply store and cost very little.

The light-weight clay aggregate may be purchased from a garden center. It is generally sold as a landscaping product.

Irrigation cycles will need adjusting as the plants grow, with more frequent cycles as the plants mature and produce fruit. Since the expanded clay is fairly porous, it retains adequate moisture for several hours once moistened. Therefore, watering during the daylight period once every two hours should be adequate. Watch your plants. If they wilt, increase the frequency of irrigation.

It is beneficial to the plants to provide oxygenation to the nutrient solution by placing an airstone in the tank, with air forced through it by an aquarium air pump via a polyvinyl hose. These materials are available from pet stores, if not from your hydroponic supply house.

This size of hydroponic growing system has sufficient space and nutrients for up to eight tomato plants (two rows of four plants each).

Space the plants 16 inches between the rows and 12 inches within the rows. However, as this high density of planting does not allow enough light to enter the crop canopy, you need to stake the plants outward as discussed earlier. The upper stems should be supported at 16-inch spacing within the rows. Alternatively, you may place only six plants in the system at 16-inch centers within the rows starting five inches in from the tray edge. The distance between the rows remains at 16 inches, beginning at four inches from the tray sides.

Sterilization of the expanded clay between crops may be done by placing it in your oven, after washing the rock. Heat it to 250 degrees F. for 20 minutes. Otherwise, simply replace it. The remainder of the system—pump, pipes, tank, growing tray, etc.—is easily sterilized by washing and pumping a 10 percent bleach solution through it. You may purchase regular household bleach from your supermarket. Dilute it in the ratio of one part bleach to nine parts water.

Nutrient-Flow Rockwool System. The nutrient-flow rockwool hydroponic system applies the same design principles as that of the aggregate system except that the growing tray is somewhat modified.

The nutrient tank can be constructed in the same way as for the aggregate system. It is possible, however, to reduce its depth as shown in Figure 14, since we are not flooding the entire growing tray as was done for the expanded clay system. The depth can be reduced to six inches allowing a solution depth of five inches which holds a volume of almost 22 U.S. gallons, which is half the volume of the aggregate system described above. A submersible pump, timer, and airstone (with an air pump) are needed to circulate the nutrient solution and maintain optimum oxygen levels as for the aggregate unit.

The connection and maintenance of solution level in the growing tray differs from the case with the aggregate system. An overflow pipe is not used. Instead, several holes at the one end of the growing tray are positioned about one-eighth inch above the bottom of the tray two to three inches in from the sides (Fig.14). The holes should be three-eighths inch in diameter. At the opposite end (the inlet end), a flexible hose of 3/8-inch to 1/2-inch inside diameter (depending upon the pump outlet fitting size), attached to the pump, enters the growing tray through a hole, just large enough in diameter to get a snug

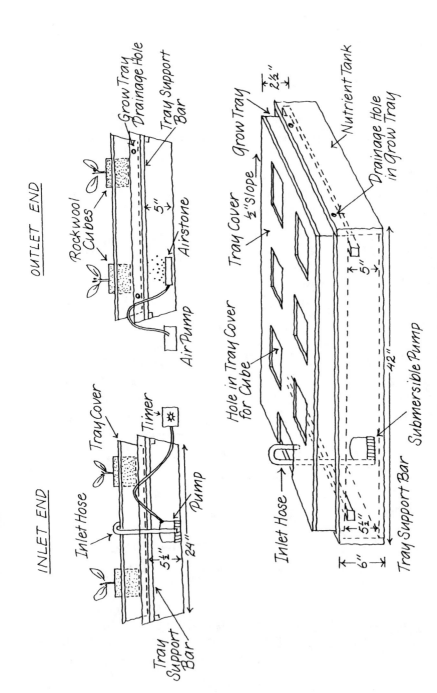

Figure 14. A nutrient flow rockwool hydroponic unit.

46

fit, located at the center of the tray one inch from the top.

Thus, the solution is pumped to the one end of the growing tray and drains at the opposite end. Slope the tray half an inch from the inlet to outlet end. This can easily be done by locating one of the support bars in the nutrient tank approximately half an inch lower (at the outlet end) than the other. Alternatively, you could simply fasten a 1/2-inch thick bar on top of the one bar at the inlet end. You get the same results. The growing tray is the same width and length as that of the aggregate culture. However, its height is less. Make it two and one-half inches high. That will be sufficient to contain the rockwool cubes in which the tomato plants are growing.

A rigid plastic cover is placed on top of the growing tray to prevent light from entering the tray so that algae and subsequent insect infestation will not develop. The cover should be white to reflect heat. It must be thick enough (at least one-eighth inch) to be opaque.

For growing tomatoes, you will have your plants growing in three-inch-square rockwool cubes inserted into the cover of the growing trays. (See Appendix for suppliers list.) These cubes must sit down through the tray cover onto the growing tray bottom.

Square holes, 3-1/4 by 3-1/4 inches, are cut in the tray cover at the positions where the plants will be located (Fig.14).

Before doing this, decide on whether you wish to grow six or eight plants in your system based upon how you wish to train your plants as described earlier. Of course, you can make more than one tray cover having different spacing and experiment with them to determine which you prefer.

Irrigation cycles for the rockwool system will be more frequent than those of the aggregate system due to the rockwool's high porosity (about 96 percent). Irrigate at least once an hour during daylight and several times during the night. With the use of the inexpensive timer the irrigation periods during any cycle will be 15 minutes, and this is adequate.

Sterilization between crops can be done by placing moist but well-drained cubes in the oven at 200 degrees F. for one-half hour. Be sure to remove the plastic side wrapper prior to heating. However, since the cost of rockwool cubes is only pennies each, there is no reason to sterilize and reuse your old ones as it may cost more in heating than the value of the cubes!

Drip Irrigation Rockwool Culture. Of all of these hydroponic units, I prefer the drip irrigation system of rockwool culture. The nutrient tank is constructed the same as for the nutrient flow rockwool system. Similarly, a submersible pump with a capacity of 80 gallons per hour is adequate. A timer and airstone (with air pump) are also necessary components. Both support bars in the tank are located one inch down from the top as the growing tray does not have to be sloped as shown in Figure 15.

The growing tray differs somewhat from that of the nutrient-flow rockwool system. Its length and width are the same, but, the height must be

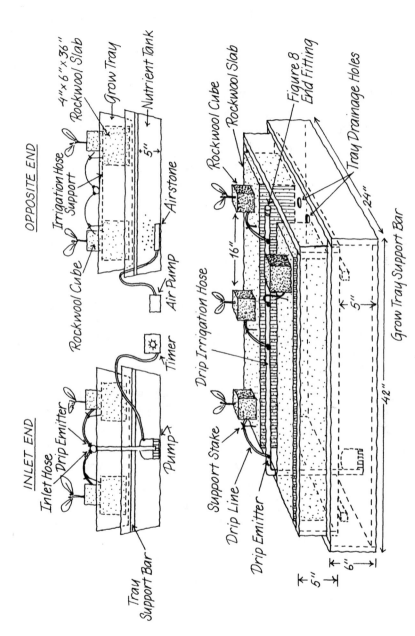

Figure 15. A drip rockwool hydroponic system.

49

greater to accommodate the 4-inch-thick rock-wool slab. A height of five inches will prevent any solution loss by unexpected runoff at the emitter location. The tray bottom requires drainage holes of 1/4-inch diameter along the center of the tray and at the ends (Fig. 15).

Drill holes at eight-inch centers along the middle of the tray and several within one inch of the ends. The pump is connected to a black polyethylene header line which enters the growing tray on one end in the center about three inches above the tray bottom. Or, it can be positioned to enter over the top edge of the growing tray supported with cross members sitting on top of the rockwool slabs. You may use short (16 inch length) wooden dowels or stakes of 1/4-inch diameter for the supports. Place one across the tops of the slabs every 12 inches for a total of three supports. The header is positioned along the center of the tray for its entire length (Fig. 15). The end is closed using a figure-eight fitting; or, you may simply bend it over and slip a three-inch piece of PVC pipe one inch in diameter over it to hold it together.

One drip emitter with a volume of one-half gallon per hour is placed at each plant location. A small spaghetti line (about one-fourth inch outside diameter) runs from the emitter to the base of the plant. These lines should be about 16 inches long.

The end of the spaghetti line is positioned on the rockwool slab by a small stake as shown in Figure 15.

These products are available from irrigation

equipment and supply stores as well as from stores specializing in hydroponics (see Appendix).

The tomatoes are seeded into the three- by three-inch rockwool cubes and later transplanted, after three to four weeks, onto the rockwool slab. The best slabs to use for tomatoes are those measuring 3 x 8 x 36 inches, or 4 x 6 x 36 inches. The slabs are wrapped with a white polyethylene plastic film. Cut holes in the top of the plastic wrapper at the locations of the rockwool cubes that will contain your transplants.

Before placing the transplants onto the slabs, soak the slabs for 24 hours with the nutrient solution to thoroughly and uniformly moisten the slab. Otherwise, dry spots may develop later, which will restrict growth of the tomatoes. Do not cut slits in the slab wrapper until the soaking process is completed. Then, cut a one-inch slit diagonally between each pair of plants on the slab on the inside surface toward the middle of the growing tray where the drainage holes are located.

Place the tomato plants, growing in the rockwool cubes, on top of the slab where you made the holes.

During the first week of growth, until the plants root into the slab, place the drip irrigation line with its stake at the top of the cube, but slightly away from the base of the plant so that moisture is not falling directly onto the crown of the plant (area where root and stem join). Then, relocate the stakes to the top of the slab, close to the base of the rockwool cube. Be careful that the plastic wrapper is cut back far enough to allow the solution to drip onto the rockwool slab itself.

No tray cover is required with this drip irrigation system. However, a white polyethylene strip placed between the plants and over the slabs to the edges of the growing tray will be useful in controlling algae growth.

An alternative to using rockwool slabs is to use six-inch-square pots filled with loose rockwool, vermiculite or a peatlite medium as discussed earlier. The seedlings are started in small rockwool propagation cubes (1-1/2 inches by 1-1/2 inches), or "Oasis Horticubes." The hydroponic system would be exactly the same as for rockwool slabs using a drip irrigation system.

Irrigation cycles need to be adjusted with the plant stage of growth and light conditions. Initially, start with one every other hour during the morning, increasing it to once per hour during the afternoon. This especially applies if you locate your hydroponic unit on a sunny patio.

Sterilization, once again, can be done by heating to 200 degrees F. for one-half hour in your kitchen stove oven. Be sure to remove the plastic wrapper first! While it is more economical to sterilize the slabs, better results may be obtained from your next crop by using new ones since there will be no structural breakdown and no accumulation of salts or plant roots. The slabs cost several dollars each at a hydroponic supply store. If you are willing to purchase at least a case of them, they can be bought for quite a bit less at a greenhouse supplier—or perhaps also at your hydroponics supplier (see Appendix).

Nutrient Film Technique (NFT)

The nutrient film technique (NFT) is a water culture system. The concept is to have a thin film of nutrient solution running past the plant roots.

As long as the solution remains very shallow, oxygenation to the roots is optimum. The upper surface of the roots is exposed to air containing 100 percent relative humidity. This keeps the roots from drying and at the same times provides optimum oxygen exchange. The plants also receive oxygen via the roots from the nutrient solution. That is why it is important to provide aeration to the solution in the nutrient tank as was described earlier using an airstone and air pump.

The simplest method of NFT, first developed by Dr. Allen Cooper in England, was to use 6 mil black, layflat polyethylene plastic. Layflat is a term describing polyethylene that is continuous in its width; that is, it forms a tube when opened up.

Layflat plastic of 10 inches in width is most suitable for the growing of tomatoes. Use a white-on-black polyethylene with the white side outside to reflect heat from your light source. It is easier to use a strip of this white-on-black polyethylene 20 inches wide and fold it up into a tent-like form (Fig. 16).

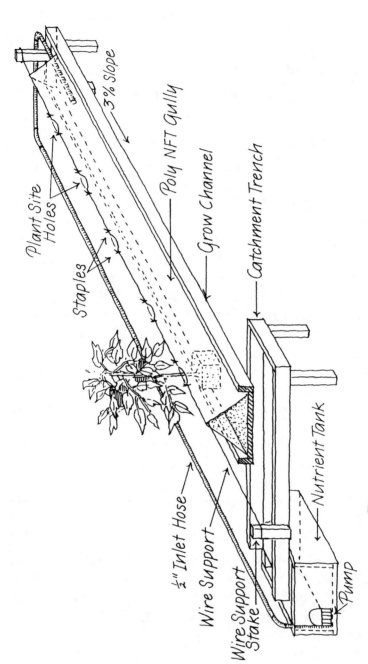

3% Slope

Plant Site Holes

Staples

Poly NFT Gully

Grow Channel

Catchment Trench

½" Inlet Hose

Wire Support

Wire Support Stake

Nutrient Tank

Pump

Figure 16. Basic NFT (nutrient film technique) system.

If black-on-white polyethylene is unavailable, try to purchase 6 mil white polyethylene which will be opaque to light.

First, stretch a wire or string, very taut, four inches above the length of bed you wish to construct. The NFT channel is to slope three percent from the inlet to outlet end. So, if you construct a bed 10 feet long, it requires a slope of: three percent x 120 inches = 3-1/2 inches.

The bottom of the NFT channel must be level across and flat, without any depressions that would retain solution. Any solution not free-flowing will soon be depleted of oxygen and cause sub-optimal oxygen levels to the plant roots. To accomplish an even slope from one end to the other of the NFT channel, you need a rigid support under the NFT polyethylene gully. Use a piece of 1- by 10-inch lumber, and nail or screw and glue 1- x 2-inch sides onto it. This will give it rigidity, in addition to preventing the polyethylene gully from moving to the side (Fig.16). Slope the channel by blocking it underneath, across the length of it, every two feet.

Place the polyethylene gully on top of the channel and staple the sides together above the wire or string support. The support wire or string could be attached to an extension of the end base blocks. Alternatively, the support wire can be replaced by use of stakes or dowels positioned across the grow-channels at 18-inch centers along the channel. These stakes pass through the NFT gully top edge to support it, thereby eliminating the need for the wire.

The sides of the grow-channels must be wider

in order to support the NFT gully. This is accomplished by use of a 1- x 4-inch piece of lumber instead of a 1" x 2" as used with the wire support system. A piece of wood can be nailed to the inlet end of the channel to prevent any solution from flowing back out of the gully.

The outlet end of the NFT poly gully should extend one-fourth to one-half inch past the end of the grow-channel to prevent the solution from running back underneath the grow-channel (Fig. 19). An anti-backflow strip could be attached to the end of the grow-channel, directly underneath the NFT gully extension. Silicone rubber sealant under this strip will prevent the nutrient solution from flowing between it and the channel.

The other components of an NFT system include: a catchment trench (if more than one channel is constructed), a submersible pump, 3/4-inch PVC pipe from the pump to the inlet end of the gullies, nutrient tank, and small diameter (one-fourth inch) black drip lines from the header pipe to each gully inlet end. Often with a number of gullies it is better to use half-inch black poly hose with a plastic valve to each inlet end from the 3/4-inch header pipe (Fig. 17).

The catchment trough or collection pipe receives the outflow of nutrient solution from the gullies, when there are more than one, and conducts it back to the nutrient tank.

The NFT channels must be raised sufficiently so that the lower outlet end projects into the catchment pipe allowing the solution to flow into it. The catchment trough may be constructed of wood lined with 6 mil black polyethylene or,

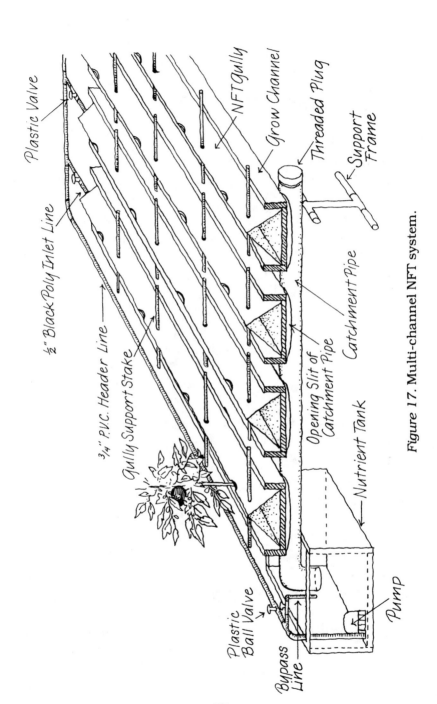

Plastic Valve

NFT gully

Grow Channel

Threaded Plug

Support Frame

½" Black Poly Inlet Line

¾" P.V.C. Header Line

Gully Support Stake

Opening Slit of Catchment Pipe

Catchment Pipe

Nutrient Tank

Plastic Ball Valve

Bypass Line

Pump

Figure 17. Multi-channel NFT system.

57

preferably, a 4-inch diameter PVC pipe. Where the NFT channels enter the catchment pipe, cut the top one-third of the pipe off with a jig saw (Fig.17). The width of the cut should be just enough to allow the NFT channel to enter the pipe.

Light can be prevented from entering the catchment pipe at this junction of the NFT channel by stapling a piece of white polyethylene to the top edges of the NFT channel and extending across the catchment pipe. The catchment pipe can be level or slightly off-level, sloping toward the nutrient tank.

The nutrient tank can be rigid plastic or of wooden construction lined with 20 mil vinyl or double 6 mil black polyethylene. Vinyl is superior in durability as it does not puncture as easily as does polyethylene. Swimming pool vinyl is suitable for this purpose.

The size of the nutrient tank is a function of the number of tomato plants to be grown. As discussed earlier in the "grow-bag" section, you need about two quarts of solution per plant per day over a five-day period.

Therefore, if your channels are 10 feet long and you have two of them, the number of tomato plants that may be grown at 16-inch centers within each row is 120"/16" = 7. The total plants in both rows then is 14, so you will construct your nutrient tank to hold a minimum of 14 x 5 x 1/2 gallons = 35 gallons. To allow 15 percent extra volume, since the tank cannot be filled completely, make a 40-gallon tank. This is equivalent to: 40 gallons/7.48 gallons per cubic foot = 5.4 cubic feet.

If you are building the hydroponic unit for use on a patio, or inside under artificial lighting, it is better to make the tank dimensions lower in height and wider and longer than you would if you could bury the tank—as you would do if it were located in your backyard garden.

A tank with dimensions of one foot high by two feet wide by three feet long would provide six cubic feet of volume. This exceeds the minimum needed by about 25 percent.

The NFT channels in this case could rest on the edge of the nutrient tank and therefore eliminate the need for a catchment trench.

With a tank three feet long, two NFT channels at 16 inches apart (center to center) will fit directly over the tank. The tank edge, if rigid enough, may support the lower ends of the NFT channels. This height of the tank will be the level of the outlet ends of the channels. Then three support legs need to be built along the length of the channels at approximately 30-inch centers. The best way to do this is to use two-inch schedule 40 (thick-walled) PVC plastic pipe as shown in Figure 18.

PVC fittings such as tees and 90 degree elbows can be used to construct an aesthetically pleasing frame. The frame should be 28 inches wide on the inside to fit two NFT channels of 12-inch width placed 16 inches apart center to center.

A submersible pump of 300 gallons per hour capacity with a pumping height of 12 feet is adequate. Since the NFT system requires continuous flow of nutrient solution, no timer is necessary. However, I would suggest you pur-

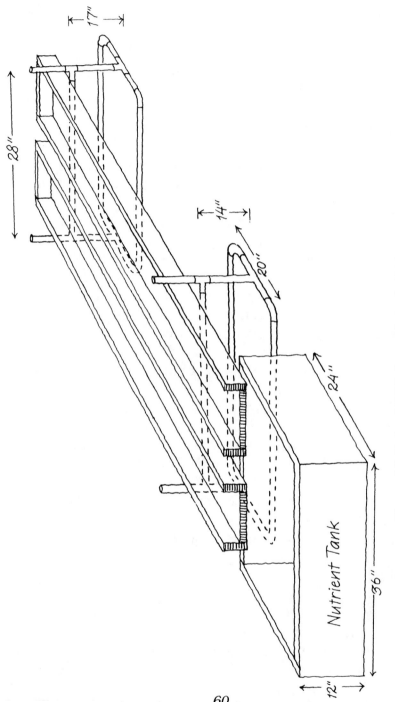

Figure 18. Raised NFT channels with PVC pipe supports (middle support not shown, for clarity).

60

chase a spare pump as a replacement, because mature tomato plants will not survive more than several hours without the flowing solution.

One method of obtaining better lateral movement of solution and more moisture retention, (should a failure in power or the pump occur) is to use a capillary matting, placed underneath the plants the entire length of the NFT channel (Fig. 19). Capillary matting is a felt-like material one-eighth inch thick which spreads water laterally like a blotting paper. It is very useful for the plants during transplanting because it distributes the solution across the entire bottom of the NFT channel preventing any running of the solution to one side of the plant, which would cause drying of the plant and possible death if not detected early. Such capillary matting is

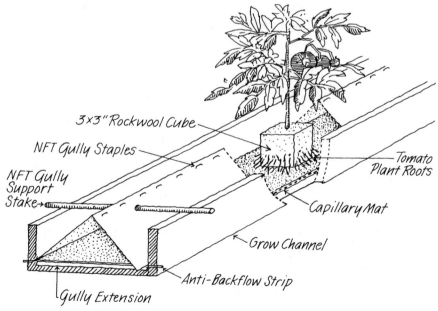

Figure 19. NFT channel and gully with capillary mat.

available from greenhouse suppliers because it is used for the growing of plants in pots on benches (see Appendix).

The solution distribution system from the pump to the inlet end of the NFT channels is made from 3/4-inch diameter PVC schedule 40 plastic pipe (Fig.20). This main runs between the NFT channels on top of the support frames to the inlet end where it has a tee with lines running to either side slightly past the center of each channel, or it can be located to one side of the grow-channels as shown in Figure 20. Glue an end cap at one end of the header and a female adapter with a threaded plug at the other end so the pipe can be cleaned periodically.

Using a quarter-inch black, flexible drip line, about one foot long, as the inlet tube, drill a hole in the PVC header. Make the hole one sixty-fourth inch smaller in diameter than the inlet tube to obtain a tight fit that will seal when the tube is pushed in. Cut the end of the inlet tube on an angle so that it will enter the hole easier. A pruning shears works well for cutting this type of tubing. Push the tube half way into the header pipe, then, apply silicone rubber sealant around it near the header pipe before inserting it the rest of the way. This will assure a good seal as a small amount of silicone sealant enters the header pipe.

Install a bypass pipe of 3/4-inch diameter PVC with a plastic ball valve to regulate the flow. Locate the bypass about one foot above the pump outlet so that the valve will be slightly above the nutrient tank (Fig.20). Place a second ball valve a few inches downstream from the

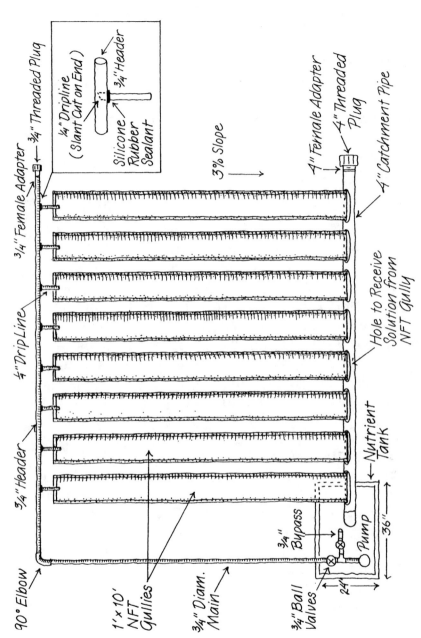

Figure 20. NFT system—piping details.

63

bypass. With the two valves you can adjust the flow of nutrient solution to the NFT channels.

The nutrient tank requires a cover to prevent light from entering it and causing algae growth. Make the cover of 3/8-inch to 1/2-inch plywood. Give it several coats of exterior latex and enamel white paint.

You will have to make notches in the cover where the NFT channels enter and where the pump lines are located. In this way it can easily be removed for changing the nutrient solution and cleaning the tank.

Sterilization of the nutrient tank during each solution makeup every other week will help to prevent disease problems. Sterilize with a 10 percent bleach solution. Be sure to close the valve downstream from the bypass so that no bleach solution enters the growing channels! That would be curtains for the plants! Rinse the tank thoroughly with raw water to remove any residual bleach solution before making up the nutrient solution.

Between crops, sterilize the entire system with a 10 percent bleach solution. Run it through all the irrigation lines and the NFT channels after removing the plants and capillary matting. Even the capillary matting can be cleaned using this bleach solution. Do not reuse the capillary matting until it is rinsed with plain water and allowed to dry. Flush the entire system with plain water after bleaching, but first allow the bleach solution to remain in the system for half an hour. Clean the top of the polyethylene gullies with a cloth and bleach solution. Fungal spores and insect eggs can accumulate on the outside of the plastic gullies. Cleanliness is very important in your future success with subsequent crops.

Tomato seedlings are grown in three-inch by three-inch rockwool cubes. After four weeks or so, transplant the plant and cube into the NFT system.

The rockwool cube will reduce transplant shock and help maintain sufficient moisture levels to feed the roots while they are starting to grow out of the blocks into the capillary matting of the channel.

NFT Pipe System

Tomatoes may be grown successfully in a somewhat different setup of the NFT system using 4-inch diameter PVC plastic irrigation pipe (Fig.21).

The components of this system are very similar to those of the basic NFT system previously discussed. The key difference is that instead of constructing growing channels and using 6 mil polyethylene for the gullies, the PVC pipe is used to serve both functions of the channels and gullies (Fig.22). This type of NFT system, as shown in Figures 23 and 24, is available from hydroponic supply stores (see Appendix.)

Drill 3-1/2-inch diameter holes in the 4-inch diameter pipe at 16-inch spacing where the plants are to be located. Use a hole saw drill, available from most hardware or building stores. Be careful that all the holes line up on top of the pipe. Make a straight line with a black felt pen

Figure 21. Tomatoes growing in a 4-inch diameter PVC
pipe NFT system.

and mark where the hole locations should be
along this line before drilling the holes.

For simplicity of comparison of this method
with the previous NFT system, we will make the
beds 10 feet long as before. A difference in height
of 3-1/2 inches from one end to the other, with
the higher end being the inlet end, will give us a
three percent slope.

As previously, construct two beds, each having seven plant holes, so a total of 14 tomato plants can be grown in the system.

The grow-pipes can be raised to about one-inch above the nutrient tank on the lower end and should be 3-1/2 inches higher at the inlet end.

This is done best with three 2-inch diameter PVC pipe supports (Fig.22) placed within one foot of each end and one in the middle. Only one support frame is shown in Figure 22 for clarity of design. The inside width of the pipe supports is 20 inches (16 inches between plant rows and twice the 2-inch radius of each pipe, as there are two pipes). If you have more than two beds the inside width is a function of the number of rows plus two inches beyond the outside rows.

A stabilizing base should be part of each support. Construct it using a tee from each vertical piece to the horizontal member and then one 90-degree elbow on each corner attached to a horizontal pipe on each side, forming a rectangular base as shown in Figures 18 and 22. Attach the grow-bed pipes to the support pipe cross-members with galvanized iron or copper pipe strapping.

If the nutrient tank is one foot high by two feet wide by three feet long, it can be placed under the outlet ends of the NFT pipes. In this way, no catchment pipe is necessary. However, if more than two growing pipes are constructed or the nutrient tank is constructed less than three feet long, a catchment pipe will be needed. Use a 4-inch diameter PVC pipe as the catchment pipe.

Connect the grow pipes into the catchment pipe with 4-inch tees (Fig.23). Slope the catch-

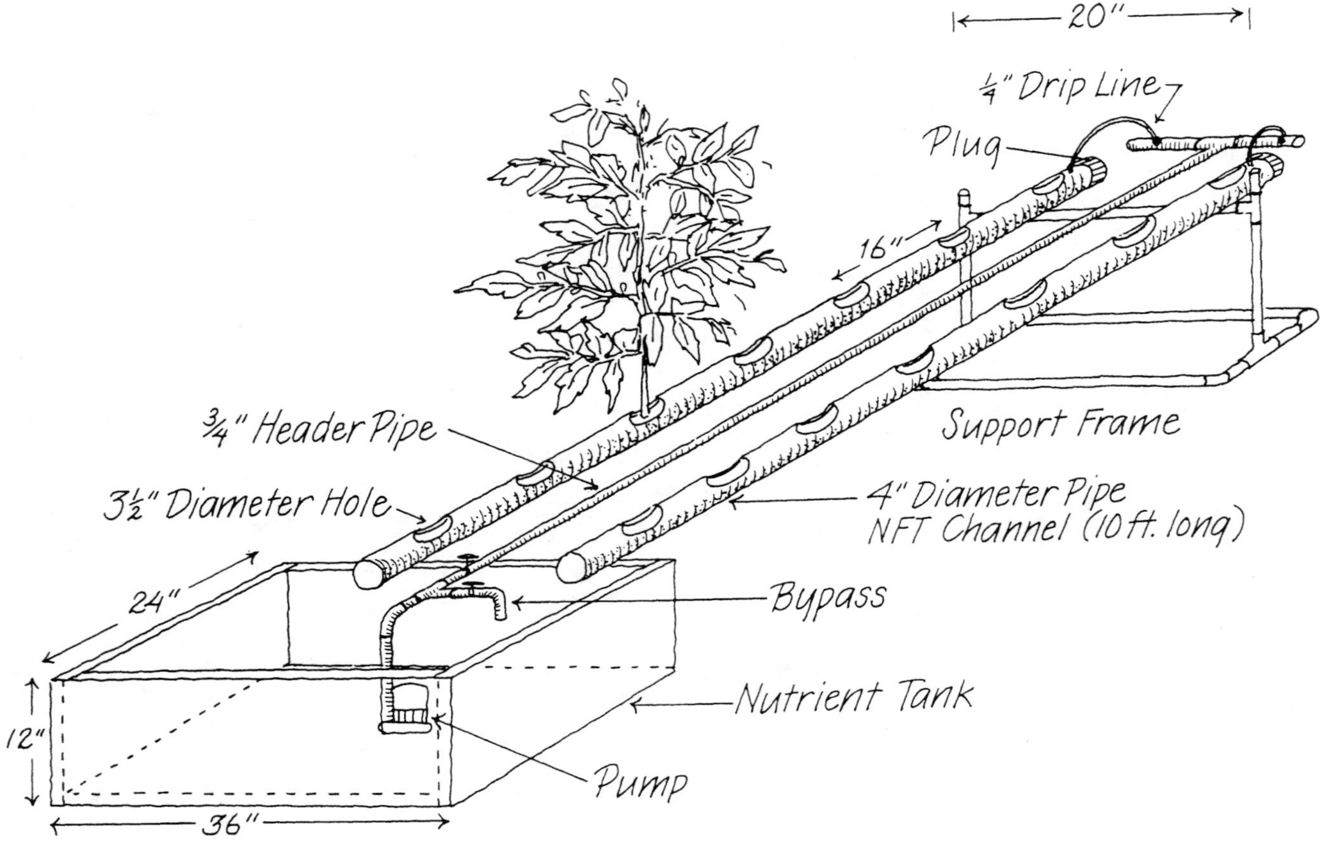

Figure 22. NFT pipe system.

ment pipe on a three percent slope toward the nutrient tank. It should enter the nutrient tank over the top edge.

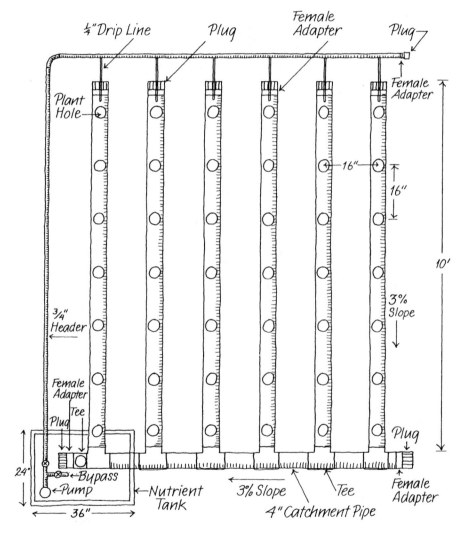

Figure 23. Multi-pipe NFT system.

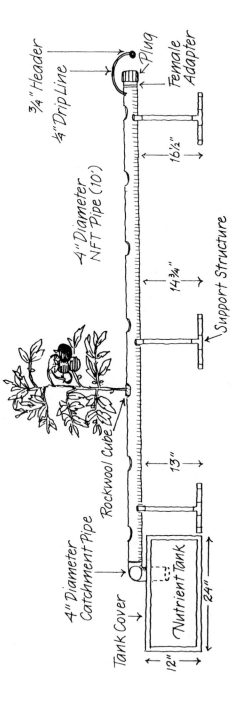

Figure 24. Multi-pipe NFT system side view showing support structures.

70

To facilitate cleaning of the catchment pipe, glue a female slip-thread adapter to each end and place a threaded plug into these fittings. Use teflon tape to seal the threads. A tee from the catchment pipe's lower end, next to the female adapter, with a short piece of pipe will pass over the edge and into the nutrient tank. Do not glue this tee, as when you wish to clean the pipe or tank or make up nutrient solutions, its removal will make easy access.

Be sure to build a cover for the nutrient tank to keep light out of the nutrient solution. A painted plywood cover is suitable. You will have to cut notches out of it to allow entrance of the catchment pipe or growing pipes, as the case may be.

Glue a female adapter with a threaded plug to the inlet end of the growing pipes. One inlet tube, a 1/4-inch diameter black drip line attached to the irrigation header, is inserted in a 1/4-inch diameter hole on top of the growing pipe downstream, next to the female adapter (Figs.22-24). One inlet line per grow pipe will provide sufficient nutrient solution as long as the pipes are not longer than 10 feet and contain no more that seven plants.

For longer beds and more plants, it would be wise to place two inlet lines per growing pipe.

The remainder of the irrigation system is similar to the conventional NFT system, using a 300-gallon-per-hour submersible pump, 3/4-inch PVC piping from the pump to the header, with a bypass and two ball valves (Fig.20). If you set up three to four growing beds, a 3/4-inch

PVC header pipe will provide adequate volume of solution flow. However, if you go to longer beds and three or more beds, it would be wise to use a 1-inch diameter PVC header pipe because you will need two inlet tubes to each grow pipe.

Sterilization between crops is achieved by flushing the system with a 10 percent bleach solution. Remove the plugs on the ends of the beds and the catchment pipe, so that you may scrub the inside of the pipes with a brush while sterilizing. Also, wipe the outside of the pipes with this same bleach solution to kill any fungal spores and insect eggs, just in case they are present. Rinse the system with plain water afterwards as was outlined for the conventional NFT system.

As with the preceding NFT system, start the tomato plants by seeding into three-inch by three-inch rockwool cubes—one seed per cube. After four weeks, transplant the plant and cube directly into the NFT pipe system. Be sure that the rockwool cube sits low enough on the bottom of the pipe so that the nutrient solution will make contact with its base. You may have to squeeze the rockwool cube slightly to do this and to fit it into the 3-1/2 inch hole, because the cubes are square!

Rockwool Culture

Rockwool culture is presently one of the most popular hydroponic methods for growing tomatoes and other vine crops, such as European cucumbers, melons and peppers.

Rockwool is an inert fibrous material produced by heating a mixture of limestone, volcanic rock and coke to 2000 degrees Celsius and extruding it as fine threads which are then pressed into loosely woven sheets. Having a 95 percent pore space, it has high water retention and yet drains well, allowing good air exchange to plant roots.

Rockwool culture may be set up by the home gardener using a returnable or recycle system (in which the nutrient solution is recirculated), similar to that of the conventional NFT (nutrient film technique) and grow-bag systems.

Growing channels, constructed of 1- x 10-inch boards with 1- x 2-inch sides, are lined with 6 mil black polyethylene or 20 mil vinyl swimming pool liner. Apply two coats of white exterior enamel paint to the wood prior to lining it. After placing the liner, staple the top edges and seal them with a strip of ducting tape as described earlier (Figs. 6 and 26). Block the inlet ends of the channels with a short piece of 1" x 2" to prevent any spillage of solution as it enters the channel via the slabs (Fig.25).

Growing channels 10 feet long will contain three rockwool slabs, while a seven-foot bed will hold two slabs. Remember, as before, the plants are positioned at 16-inch centers. Therefore, place three plants per 36-inch long slab. One within four inches from each end and the third in the center. Space adjoining slabs 10 to 12 inches apart.

The outlet ends will spill directly into the nutrient tank if only two 10-foot beds are

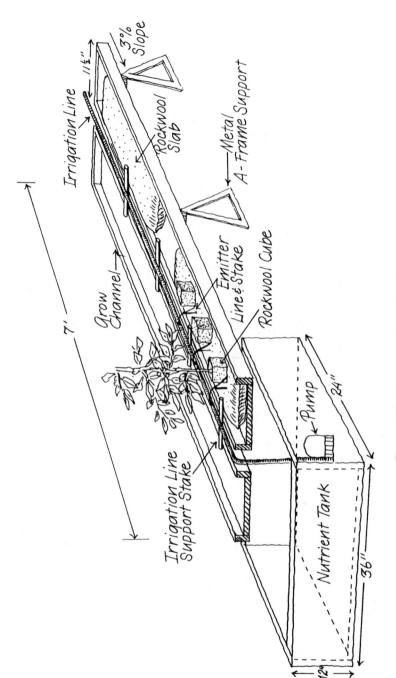

Figure 25. Basic rockwool culture system.

74

constructed and if the tank is at least three feet long as presented earlier under the NFT section. Otherwise, if the width of the channel array is greater, a catchment trough must be constructed to collect the nutrient solution and conduct it into the nutrient tank.

The growing channels must have a three percent slope toward the catchment trough or nutrient tank.

As before, the catchment trough may be constructed of four-inch diameter PVC pipe or, in a manner similar that used for the growing channels, of wood lined with 6 mil black polyethylene or 20 mil vinyl.

Such a trough should be covered with 6 mil white polyethylene to help reflect heat. It can be constructed of a 1- x 8-inch bottom with 1- x 2-inch sides and end (Figs. 5 and 7). One end will have to sit on the edge of the nutrient tank with the tank located perpendicular to the catchment trough and parallel to the beds. It is easier to use four-inch PVC pipe for the catchment pipe (trough) as outlined above with the NFT pipe system since it can easily be constructed in an "L" shape to enter the nutrient tank.

The supporting frame for the growing channels should be constructed of two-inch PVC pipe as was described for the NFT pipe system (Fig. 18). The support frames could also be made of aluminum 1-1/2 inch angle and/or square tubing as shown in Figures 25 and 27.

A triangular or "A" vertical frame should hold a cross support every 30 inches across the bed length. This would particularly apply to a multi-bed hydroponic system. The triangle base is 12 to 14 inches wide. A cross member of 1-1/2 inch square tubing is attached to the apex of the triangle frame. Bolt all pieces together.

If the rockwool system consists of only two growing channels, three supports of similar dimensions as shown in Figure 24 for the NFT pipe system, will bear the weight of the rockwool system. The only difference in dimensions is the width. The inside width should be 26 inches, allowing 16 inches between the rows and 10 inches for the channels (two times half the width of each of the two channels).

The irrigation system consists of a 300 gallon-per-hour capacity submersible pump, timer, 3/4-inch PVC pipe main line with a bypass and two ball valves, 1/2-inch diameter black polyethylene hose running the length of the beds between the pair of channels, 1/2-gallon per hour emitters (one to each plant) and 1/4-inch drip line from each emitter to the base of each plant (about 14 inches in length) as shown in Figure 26.

The black poly hose is supported between the grow channels by stakes or dowels (1/4-inch diameter by 16 inches long) placed across the tops of the slabs at 16-inch centers along the beds.

It is important to set the emitters directly into the black poly distribution line rather than at the end of the emitter drip line. This will prevent precipitation and plugging of the emitters which can occur if the emitters dry out between irrigation cycles.

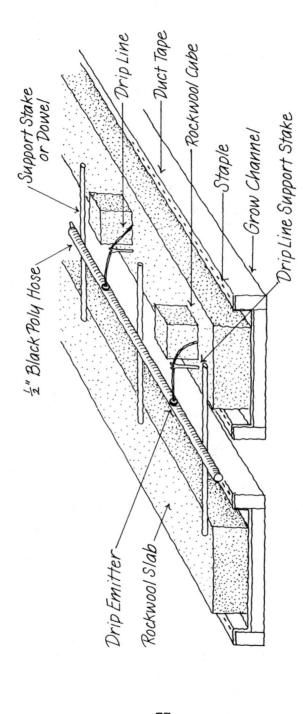

Support Stake or Dowel

Drip Line

Duct Tape

Rockwool Cube

Staple

Grow Channel

Drip Line Support Stake

½" Black Poly Hose

Drip Emitter

Rockwool Slab

Figure 26. Rockwool culture—detail of drip system.

77

You may purchase a special tool at a greenhouse or irrigation supplier that will puncture the correct size hole into the black polyethylene hose so that a complete seal is obtained. If you mistakenly punch too many holes or put them in the wrong place, you can purchase "goof plugs" with which to correct any such mistakes.

As described earlier in the section on bag culture using perlite medium, the emitter line is staked initially, after transplanting, near the base of the plant on the growing cube. After five to seven days, as the plants root into the slab, relocate the stake and drip line onto the slab at the base of the cube (Fig.26). This reduces moisture levels directly in the crown of the plant and thus will lessen chances of diseases.

Six-inch wide by 36-inch long rockwool slabs four inches thick are best suited for the growing of tomatoes. These are placed in the growing channels so that 16-inch centers exist between plants. To do this, locate one tomato plant in the center of the slab and one to either side, 14 inches apart. That places the outside plants within 4 inches of the slab ends. Allow 12 inches between slabs. You get the 16-inch spacing by bending each of the plants within four inches of the ends of the slabs, two inches toward each other and staking or stringing them up to a support wire (Fig. 27). The important point here is that the upper stems are at 16-inch centers to allow adequate light for good growth.

Sow tomato seeds into three-inch by three-inch rockwool cubes and transplant these cubes containing the tomato plants onto the rockwool

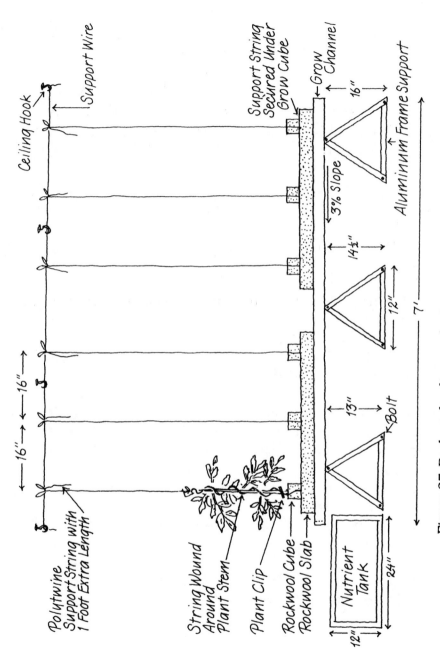

Figure 27. Rockwool culture—plant spacing and support.

79

slabs after three to four weeks of growth as described earlier in the section on small indoor units using rockwool medium.

Remember to presoak the slabs for 24 hours, using the nutrient solution, prior to cutting the drainage slits on the inside face of the slab near the irrigation line as described for the "rockwool hydroponic unit."

One handy tip: if you are going to support your plants with a string and overhead wire, place the end of the string between the base of the rockwool transplant cube and the top surface of the slab when transplanting. This will secure the lower end of the support string.

Wind the string in a clockwise direction around the stem of the tomato plant. Use a plastic plant clip under the first leaf to secure the string. These procedures are discussed later in the chapter on cultural practices.

Irrigation cycles should be every hour during the daylight and several times during the night. A 24-hour timer in series with a 60-minute timer is needed to get proper irrigation cycles and duration.

The duration of each irrigation cycle should be sufficient to allow at least a 20 percent runoff of solution through the slab. This will leach the slabs preventing any build-up of nutrient salts. Do not be concerned about the amount of runoff as the solution is returned to the nutrient tank, it is not wasted! The pH and electrical conductivity (EC) should be monitored daily. Maintain the pH between 6.0 and 6.5 and the EC between 2.2 and 3.5 milliMhos. Using a syringe, take several

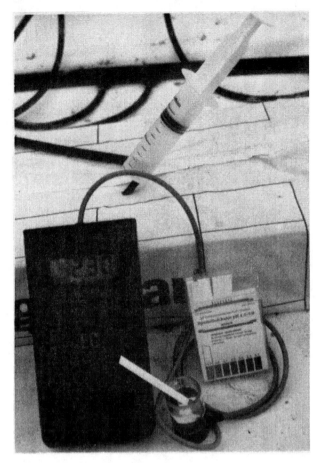

Figure 28. Monitoring of slab—using a syringe, EC meter and pH paper.

solution samples from the middle of the slab between the plants (Fig.28). Details of pH and EC will be discussed in the section on nutrients.

Sterilization of the channels, nutrient tank, irrigation lines and catchment trough can be done with a 10 percent bleach (sodium hypochlorite) solution. The rockwool slabs may be sterilized by heating them for one-half hour at 200 degrees F. in your kitchen stove oven—after removing the plastic wrapper.

They can be sterilized several times before any structural damage may occur that could reduce their porosity, which would reduce oxygen up-take to the plant roots. You can save money by sterilizing the slabs (at a few dollars each), depending on the volume you buy and where you buy them (see Appendix for suppliers).

Chapter 2

The Nutrient Solution

In a hydroponic system, the plants are fully dependent upon you to give them all of the nutrients they need for a healthy life. Unlike growing plants in soil, which provides nutrients to the plants but not necessarily in the correct amounts, a soilless system will, if properly formulated, offer the plants optimum levels of all their essential elements.

Essential Elements

Plants require sixteen essential elements. These are the basic elements that plants must assimilate in order to complete their life cycle.

Thirteen of these essential elements are minerals normally available to plants from the soil as they are released into the water solution that may exist in the soil. Often, however, due to inadequate supplies or unavailability, the lack of any one or several will greatly restrict plant growth and development.

83

In a hydroponic system, all thirteen of these elements must be put into the nutrient solution.

The essential elements for plants are: nitrogen (N), phosphorus (P), potassium (K), calcium (Ca), magnesium (Mg), sulfur (S), iron (Fe), manganese (Mn), zinc (Zn), copper (Cu), boron (B), molybdenum (Mo) and chlorine (Cl).

The other three elements plants require are: carbon (C), hydrogen (H) and oxygen (O). The sources of these are air and water. Plants assimilate these through photosynthesis, using energy from sunlight or artificial lighting.

The uptake of these plant foods is the same whether from soil or from a soilless medium, as in a hydroponic system.

In the soil, the elements become available through weathering of minerals such as rock, sand, silt and clay and through decomposition of organic matter into its basic elemental constituents. These basic minerals adhere to the soil particles and are released into the soil water as ions (individual charged atoms) which enter plants roots immediately in contact with them. This occurs in a process called ion exchange in which certain ions are released by the plant roots in order to take up the other essential element ions. The process is the same in a hydroponic system.

In soilless systems, hydroponics, the nutrient solution contains all of the essential elements in ionic form. As the plant roots are constantly in contact with this solution, uptake of these ions by the plant takes place continuously.

The difference between soil and soilless systems is that in soil there are always some residual elements present; in a hydroponic system, using an inert medium and raw water having very little, if any, nutrients, all of these plant essential nutrients must be added to the nutrient solution.

Your job, as the home gardener, is to add all of these essential plant minerals in the right proportions to provide the plants' diet! If you do this correctly, your plants will get all of their nutrients *at optimum levels.* This, together with the right environmental conditions such as light and temperature, will enable your plants to thrive and produce high yields of fruit.

Be aware of one fact—you are in control! If you leave out one or more of the essential elements or add too much of any one of them, you may damage your plants. So, if you are making up your own nutrient blends, follow the formulations carefully to avoid problems.

Instead of mixing your own, you may purchase complete hydroponic mixes for your tomato plants from garden centers and hydroponic suppliers (see Appendix). Be accurate in weighing out these nutrients for use in your nutrient solution. You may need a gram scale or balance to do this.

I think it is more satisfying to make up your own nutrient mixes. As you come to understand your plants' needs at various stages of growth and under different environmental conditions, especially light intensity and day-length, you can alter your nutrient formulation somewhat to adjust for these changes in growth or environmental conditions.

The objective is to provide optimum levels of nutrients for the plant at all times. Your tomato vines will reward you by producing more fruit and better quality fruit in both appearance and taste!

pH:

An understanding of plant nutrition, pH and electrical conductivity (EC) will greatly assist you in providing optimum care of your tomato plants.

The pH is a measure of acidity or alkalinity of a solution. The pH scale runs from 0 to 14 (acid to base). A pH of 7 is neutral, below 7 is acid and above 7 is basic. The optimum pH for tomatoes is between 6.3 and 6.5. As pH shifts beyond the optimum range, the availability of some of the essential elements to root uptake is reduced.

One pH unit shift, say from 7 to 6 or 6 to 7, represents a ten-times shift in acidity or alkalinity. Thus, a full unit shift in pH, away from the optimum range, will result in a significant decrease in nutrient uptake by plant roots. For this reason, it is important to monitor pH of the nutrient solution and maintain it within the optimum range.

pH Adjustment

You will have to adjust pH in one of two directions: upward, if it is below 6.3; or downward, if above 6.5.

To increase the pH (alkalinity), you must add a base to the nutrient solution. To do this, use potassium hydroxide (KOH), sodium hydroxide (NaOH) or bicarbonate of soda ($NaHCO_3$). The latter, baking soda, is the safest to use, unlike the other two, which burn your skin. If you use the hydroxides, *wear gloves and safety goggles.* Follow the supplier's safety instructions.

To decrease the pH of the nutrient solution (make it more acidic), add an acid. Some acids that are used are: phosphoric acid (H_3PO_4); sulfuric acid (H_2SO_4)—battery acid; nitric acid (HNO_3); hydrochloric acid (HCl)—muriatic acid, commonly used in swimming pools; or acetic acid, as vinegar.

Be very careful with acids. I do not recommend using sulfuric acid or nitric acid. They are especially dangerous and will burn your skin severely. If you decide to use them anyway, *always wear rubber gloves and eye goggles—and be careful to follow the supplier's safety instructions.* Fumes from these two acids, especially, as well as from muriatic acid, are very toxic, so *you should also wear a respirator* rated for adequate protection from these specific acid fumes.

While acetic acid is the safest to handle, it is not as strong as the others, so it may be more practical and effective to use muriatic acid, *excercising care and observing all the instructions and cautions on the container and from the supplier.*

Remember, always *add acid to water, never the opposite* as it will heat up and splash! Add the acid directly to the nutrient solution, a little at a time, testing the pH with litmus paper after each addition.

One of the better indicator papers tests a range of pH from 4.0 to 7.0 in 0.3 to 0.4 increments showing specific color changes from yellow (4.0) to dark blue (7.0). This particular indicator paper is called "colorpHast," made by Merck. It is available from your hydroponics supplier or from greenhouse suppliers listed in the Appendix.

You may wish to purchase a pH meter, but these are expensive ($300 to $500) and are very delicate. They also need frequent calibration with standard test solutions to remain accurate. The indicator paper is quite inexpensive.

Electrical Conductivity (EC)

Fertilizer salts, when dissolved in water, dissociate into individual electrically charged units called ions. For example, KNO_3 forms K^+ and NO_3^- ions in solution. Plant roots, when in contact with these elements in their ionic state, absorb them from the nutrient solution.

Ions in solution conduct electricity. This conductance of electric current in solution can be measured by an instrument called an electrical conductivity meter. These meters are available as small pocket models costing less than $100. The more accurate and reliable ones cost several times as much and are available from greenhouse and hydroponic suppliers.

Electrical *conductance* is measured as "Mhos." Since the charge is very small, the EC meters detect "milliMhos/cm" which is a measure of conductance through one cubic centimeter of solution. Electrical *conductivity* measures the electrical conductance of the total dissolved solutes in the solution. It does not indicate the level of any individual ion. To maintain the accuracy of your EC meter, calibrate it weekly with a standard test solution which may be purchased from greenhouse suppliers.

If you wish to make your own standard test solution you may do so in the following manner: For a solution having an EC of 1 milliMho, add 1 gram of calcium nitrate to 1 liter of distilled water. For an EC of 2 milliMho, add 2 grams of calcium nitrate to 1 liter of distilled water.

The optimum range of nutrient solution electrical conductivity for tomatoes is 2.0 to 3.5 milliMhos.

Vegetative growth of tomatoes can be slowed by raising the EC, while lowering the EC will stimulate rapid nutrient uptake and increased vegetative growth.

Therefore, generally during dull weather (poor light conditions), the EC should be increased as should also be done during fruiting to promote fruit development more than vegetative growth. Addition of any fertilizer salts will cause an increase in EC. To avoid increasing nitrogen levels excessively, add monopotassium phosphate (KH_2PO_4) or potassium sulfate (K_2SO_4) to increase the solution electrical conductivity.

While EC measures the total solutes (minerals dissolved) in the nutrient solution, it does not indicate the amounts of any specific element. Therefore, even if the EC of a nutrient solution appears to be within the optimum range, the proportion of each specific nutrient element may not be optimum, so you must still change the nutrient solution on a regular basis unless you have a complete analysis done by submitting a solution sample to an analytical laboratory (see Appendix).

Water Quality

Most household tap water contains some nutrient elements. If you live in a dry area such as southern California, Arizona, etc. your water will most likely be high in magnesium and calcium in the form of carbonates (magnesium carbonate and calcium carbonate).

Many areas have waters high in sodium chloride. For instance, in Ventura county of California, where I have experience in using various well waters for the growing of herbs hydroponically, the sodium and chloride levels range from 50 parts per million (ppm) to in excess of 150 ppm during the rainy to dry seasons, respectively.

These high levels of sodium chloride in the raw water, if used in a circulating hydroponic system will build up in the nutrient solution to toxic levels for tomatoes since the plants do not take it up as a nutrient.

Each time you add water to your system, you are adding more sodium chloride until, over a period of two to three weeks, levels may be in excess of 300 ppm. Through solution analyses, I have found this to be the case in the growing of basil in a recycling NFT system.

The magnesium and calcium carbonates are not a problem, since these are essential elements, and the carbonates help stabilize the solution pH. However, you will need an analysis of your raw water, to know what levels already exist, before making up a nutrient formulation.

In some cases this water analysis may be available from your county water department. If not, you may have an analysis done at a testing laboratory (see Appendix). The cost for a complete analysis at a university laboratory is approximately $35.

Once you receive the results, simply subtract the raw water levels of these elements from the amounts needed in your final formulation and add only the difference. In some areas, again as in Ventura county of California, there is sufficient boron in the water due to high boron mineral deposits within the water source region. If your water contains levels of boron equal to or in excess of what your nutrient solution requires, do not add any more.

Alternatively, if your water is too high in minerals or if your hydroponic system is a smaller indoor unit, you may wish to use distilled water—available at your local supermarket. Distilled water has no elements present, so you must add all of the nutrients called for in your nutrient formulation.

Parts Per Million (ppm)

In working with nutrient formulations, it is important to understand the concept of parts per million or milligrams per liter (mg/l), which are equivalent.

One part per million is simply one part of one thing in one million parts of another. For example: 1 ppm of potassium (K) is one part of potassium in one million parts of water, which is 1 milligram of potassium in 1000 milliliters (1 liter) of solution (water): 1 mg=1/1000 gm; 1 l=1000 ml; Therefore,

1 mg/l=1/1,000,000=1 ppm.

Fertilizer Salts

The essential elements required for the nutrient solution are available in the form of fertilizer salts. Always purchase the best grades since they are most soluble. You need not use laboratory reagent grades; but, if your hydroponic system is a small indoor unit needing very small amounts of fertilizer salts and you do not have storage space, you may wish to purchase these highly purified forms, but they will be more expensive than regular fertilizers.

Table 1 in the Appendix lists the fertilizer salts to use to provide the various essential elements in your nutrient solution.

Prepared hydroponic mixes containing the elements in suitable proportions can be purchase from garden centers and hydroponic suppliers. If you have the time and wish to be able to change formulations to suit your tomato plants' needs during different growth stages and environmental conditions, it is, as I have indicated, preferable to make up your own nutrients by following the formulations and procedures presented in the Appendix.

Nutrient Formulations

Nutrient formulations are recipes of plant foods which provide optimum levels of each essential element to your tomato plants. The calculation of these nutrient formulations is not presented in this book, as they are somewhat complex.

If you wish to pursue such information, especially if you are experienced in hydroponics, or may wish to do so later, the derivation of nutrient formulations is presented in *Hydroponic Food Production*, by this author (Woodbridge Press). Other publications are listed in the Appendix. Additional information on nutrient formulations, preparation, adjustment for water quality and stock solutions is also presented in the Appendix.

Chapter 3

Cultural Practices for Growing Tomatoes

Environmental Conditions

Proper control of light and temperature at optimum levels for the growth of tomatoes will increase fruit yields to their full potential. If plant nutrition is maintained at optimum levels in a hydroponic system, but optimum environmental conditions are not, fruit production will be less than the plants are capable of yielding. Light and temperature, therefore, can be either limiting or maximizing factors.

Light

Light quality, intensity and duration affect plant growth. Tomatoes grow best under the natural, full sunlight of the summer months when day-length is at its maximum. Anything less than these light conditions can reduce productivity.

You might think then, "Why attempt to grow tomatoes indoors, when I know that these conditions cannot be achieved?" The fact is, if you grow at least four to six plants, even under suboptimal light conditions—while your per plant yield may be less than optimum—you will still produce enough excellent fruit for your personal needs.

If you are growing under less than ideal light conditions, it is even more important that plant nutrition is not restricting growth. For this reason, a hydroponic system plays a positive role in improving fruit yield through proper plant nutrition. In fact, under poor light conditions, with the adjustment of electrical conductivity of the nutrient solution, as discussed in Chapter 2, the plants' growth can be modified specifically to increase fruit production.

When growing tomatoes indoors, locate the hydroponic system in an area having the best natural light. Ideally, this would be an enclosed patio facing south. If you have no patio, the next best site is near a large window.

During the winter months, when natural light is of lower intensity and short duration (day-length of 8 hours or less), it is imperative to use supplementary artificial lighting above the plants to increase the light intensity to 1000 foot-candles at plant height and day-length up to 14 or 16 hours.

The least expensive artificial light having the correct quality (color) is high-output fluorescent. Use the dual tube units, with a reflector on the fixture. Position one double-tube unit above each row of tomato plants, supporting the units from the ceiling with "jack chains" (small chains available at any building store) and "S" hooks. In this way, you can easily raise the lights as the plants grow, keeping them about 8 to 10 inches above the growing tips of the plants.

Use a timer to turn the lights on and off. This has several advantages. It allows you to be away and, secondly, especially during the winter months, you can operate the lights over the night period when electrical rates are often less.

Since tomatoes are almost day-neutral in their day-length response, a short, dark period between lighting intervals will satisfy their dark period needs. For example, have the timer turn on the lights from 10:00 PM to 2:00 PM; they will then have had a four to five-hour dark period before 10:00 PM. As the day-length increases in the spring months and natural light intensity improves, the period of supplying supplementary lighting can be reduced.

If your plants are located in a patio having relatively good natural light, you can use the supplementary lights to extend the day-length during the early morning and late afternoon when natural light is of low intensity. For example, set the timer to turn the lights on at 4:00 AM and off at 9:00 AM, on again at 3:00 PM or 4:00 PM, and off at 6:00 PM or 8:00 PM.

Temperature

Temperature is also an important factor in plant growth. However, unlike light, temperature is much simpler to control at optimum levels. Most plants, including tomatoes, grow best under two different temperature regimes, one during the night and the other during the day. It is important to coordinate this diurnal temperature fluctuation with your lighting cycle. Any light period must be considered daytime for the plants and, therefore, must provide daytime temperatures whatever the actual time of day.

The best temperatures for mature tomato plants bearing fruit are 60-65 and 65-75 degrees Fahrenheit (F) during the night and day periods, respectively. Optimum temperatures for seed germination are 65-70 degrees F. for the day and night periods. Seedlings from 10 days to 4-5 weeks old, until transplanted, do best with optimum night and day temperatures in the range of 58-62 and 65-75 degrees F.

Temperatures may be regulated with thermostatically controlled heaters and fans. If your hydroponic system is in a closed room, temperatures may be controlled with your existing thermostatically-controlled central heating and air conditioning system.

If the plants are grown in an enclosed patio that does not receive heat from the remainder of the house or apartment, you need to provide a separate electric space heater, either with a built-in thermostat or connected to a remote thermostat.

Generally, a separate thermostat located among your plants is beneficial regardless of whether or not the heater has one.

An exhaust fan, having a capacity to exchange the total volume of air in the enclosed patio within half a minute, will maintain temperatures to within 5 degrees F. of the ambient temperature. A thermostat may be installed in series with the exhaust fan to control its operation. The thermostat should be installed inside the crop canopy so that the temperature it measures will be representative of the average temperature around the tomatoes.

During the high light-intensity summer months it is possible to reduce the light entering the patio and hence the heat build-up due to the "greenhouse effect" (the trapping of heat by a change in light wavelength from short to long as it enters the patio). Apply a shade compound to the outside of the patio glass or other covering material you may be using. Use a concentration of shade compound to give a 25 percent reduction in light.

The reduction in inside light from that outside can be determined using a light meter. Compare the light levels outside and inside the patio. Take these light level readings at mid-day when light intensity is greatest.

Shade compounds and light meters are available from greenhouse suppliers (see Appendix) or perhaps from your hydroponics supplier. Purchase a three to four-month permanency formulation so that it will be easily washed off with water (or natural rainfall) by the autumn.

The shade compound is easily applied with a paint sprayer. If you live in an apartment, be careful to apply it on a calm day, usually during the early morning, so as to avoid any drift to your neighbors or onto walls of the building!

Obviously, on hot days during the summer you will not be able to keep temperatures in the optimum range for the tomatoes. If the ambient temperature, for example, is 80 degrees F. the temperature in the patio will be 85 degrees F. or more.

Methods of cooling below ambient temperatures include an evaporative cooling pad, misting and fogging. The cost of an evaporative cooling pad and fogging system is substantial. Consequently, the only economically feasible alternative may be to increase the relative humidity with an overhead misting system or to place an ordinary humidifier in the patio.

Under high light, tomatoes are fairly tolerant of temperatures up to 90 degrees F. Beyond that, the pollen in the flower begins to dry and fruit set does not form. This will reduce the fruit yield, but, as mentioned earlier, if you have six or eight plants, they will still provide enough fruit for your own use.

Tomato Seedlings

Sow seed into rockwool propagation blocks, rockwool growing cubes, "Oasis Horticubes," peat pellets, or a peat-lite medium contained in "Com-pack" plastic trays (Fig.29). Rockwool blocks, cubes and "Horticubes" are best suited

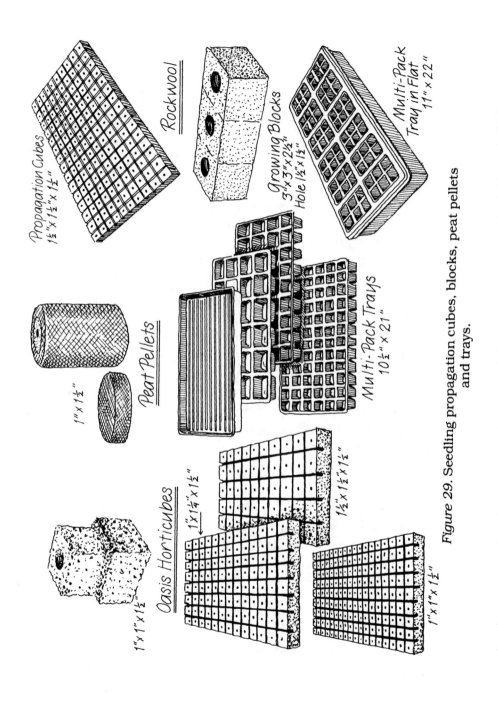

Figure 29. Seedling propagation cubes, blocks, peat pellets and trays.

Rockwool

Propagation Cubes
1½" x 1½" x 1½"

Growing Blocks
3" x 3" x 2½"
Hole 1½" x ½"

Multi-Pack
Tray in Flat
11" x 22"

Peat Pellets

1" x 1½"

Multi-Pack Trays
10½" x 21"

Oasis Horticubes

1" x 1¼" x 1½"

1½" x 1½" x 1½"

1" x 1" x 1½"

1" x 1" x 1½"

for water culture systems such as nutrient film technique (NFT), pipe NFT, rockwool, and grow-bags having a sawdust or perlite medium (see Chapter 1 for information about these systems).

Peat pellets retain too much moisture in the above types of hydroponic systems. They are suitable for use with sand, gravel or expanded clay cultures, either in beds or bags. Peat pellets and "Com-pack" trays, with a peat-lite medium, can be used for growing tomato seedlings for later transplantation into plastic pots or bags containing a peat-lite growing medium.

Tomato seed can be sown into the "smaller" rockwool propagation blocks available (they contain 98 1-1/2" cubes), then transplanted into the three-inch by three-inch wrapped rockwool cubes (Fig.30). (Remember that all of the supplies mentioned in this book are readily available from your hydroponics supplier or other suppliers listed in the Appendix.)

Before sowing seeds in the rockwool cubes, soak them with the dilute seedling nutrient solution formulation (see Table 2 in Appendix) for 24 hours to adjust the pH and assure complete wetting of the cubes. The rockwool propagation block cubes have small, recessed holes in them to hold the seed in place during watering. If you water by hand several times a day, it is helpful to prevent drying of the seed between watering by filling the hole after seeding with vermiculite or perlite.

Transplant the small cubes into the larger three-inch by three-inch cubes when the first true leaf unfolds (about two weeks old) as shown in Figure 31. As the tomato plants grow in these

Figure 30. Rockwool propagation cube (top), growing block
(center) and slab (bottom).
Courtesy of Agro-Dynamics, Inc.

larger cubes, space them apart so that the leaves
are not greatly overlapping. The spacing by the
four to five-week stage of growth should be
six-inch by six-inch, center to center of the
cubes.

After watering the seedlings, complete drainage of excess moisture is essential. If solution
collects underneath the cubes, puddling will
occur, causing root death of the plants. The
easiest procedure for good drainage of the cubes
is to place them in a flat containing drainage

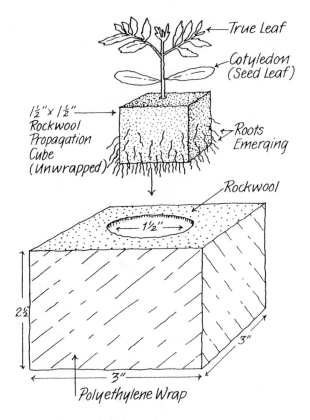

True Leaf

Cotyledon (Seed Leaf)

1½" x 1½" Rockwool Propagation Cube (Unwrapped)

Roots Emerging

Rockwool

1½"

2½"

3"

3"

Polyethylene Wrap

Figure 31. Transplanting small rockwool cube into larger growing block.

holes. This flat should be set into a second one, without holes, to collect the water. Raise the first flat slightly above the second one by placing several 1/2-inch diameter PVC pipes across the bottom of the lower flat without holes. Each day, empty the bottom tray of excess solution.

When watering the seedlings, water to at least 25 percent runoff to get complete movement of solution with its essential plant nutrients through the rockwool cubes.

"Oasis Horticubes," because of their small size (1-1/2" x 1-1/2" x 1-1/2"), are more suitable for the growing of lettuce and herbs than for tomatoes. However, if you take the right precautions, they can be used for tomatoes, but the plants will have to be transplanted at an earlier stage than if they were in rockwool cubes. After sowing the seed, cover the holes with perlite or vermiculite as you did for rockwool cubes.

Transplant at the three to four-week stage of growth before the plants become too large and dry out too frequently between watering.

"Horticubes" come as a block of 50 cubes fused together at the base. After 10 to 14 days, once the first true leaf of the tomato plants has fully expanded, separate the cubes of the block and space them to four-inch by four-inch centers. Place the "Horticubes" in several flats and water thoroughly, as discussed above for rockwool cubes, to achieve proper nutrition and drainage.

As the plants get larger, close to the 3-week stage, it may be advantageous to place the plants directly in the flat without holes so that some solution will be present at their base to prevent drying. This practice might not be optimum for plant growth, but it will prevent desiccation and death of the plants should they be depleted of water for a number of hours when you are out!

Peat-pellets are compressed discs of peat wrapped with a nylon mesh. When soaked in water for 15 minutes they expand to a round plug approximately 1-1/2 inch high. They contain fertilizers and are pH adjusted so they will carry a tomato plant for four to five weeks.

After thoroughly moistening the pellets, sow one tomato seed per pellet by pushing the seed 1/4 inch into the medium with a pencil. Place them tightly together in the flat-in-flat method described above (Page 102) for rockwool cubes and water them several times a day to prevent drying.

After 10 to 14 days respace them at four-inch centers and by four to five weeks transplant them. At that time they should have four or five fully expanded true leaves and the plants, if healthy, will be as high as wide (Fig.32).

When sowing tomatoes in "Com-pack" trays containing peat-lite vegetable mix of peat, perlite, and vermiculite, use the larger trays having 24 or 36 cells per flat (dimensions of 10-1/2 inches x 21 inches). The plant spacing in these trays is approximately 2-1/2 inches x 3-1/2 inches and 2-1/2 inches x 2-1/2 inches, respectively. This

Figure 32. Healthy tomato transplants in peat pot (left) and
in peat pellet (right) with roots emerging,
ready to plant.

spacing is fairly close for tomatoes. At best, they could be grown for three to four weeks in these trays before overlapping of the plants' leaves would cause competition for light, which would cause "legginess" should they remain at that spacing.

Transplant the tomatoes into their grow-bags before they begin to get spindly. The plants will have a firm root ball (mass of roots containing the medium) at that stage, so you can "pop" them out of the trays by pushing on the bottom while pulling the plant by its stem. For convenience, to avoid spilling water from the flats at the time of watering, place a second flat, without holes, under the one with drainage holes.

Tomato Varieties

There are a number of tomato varieties that grow vigorously in hydroponic culture. Others may be grown, but unless you use the greenhouse varieties that are especially bred for controlled environment growing, yields will be less. In addition, most of the greenhouse varieties have resistance to many of the diseases common to tomatoes.

While you may prefer to grow bush type tomatoes, the most efficient use of your hydroponic system is achieved by growing staking tomatoes. With staking varieties, the tomato plant is trained vertically and produces fruit continuously up the main stem. Training is discussed later under "Training and Pruning."

Some of the more popular tomato varieties for

hydroponic culture are: Tropic, Dombito, Caruso, Jumbo, Vision, Larma, Perfecto, and Laura. Seeds for these are available from seed companies and hydroponic suppliers (see Appendix). Try several different varieties to determine which ones do best with your specific hydroponic system and environmental conditions.

Transplanting and Spacing

In general, transplant once the tomatoes have three to four true leaves and their roots are penetrating the bottom of the growing cubes as shown in Figures 31 and 32. Handle the plants carefully by their stems. Avoid breaking leaves or scraping the skin (epidermis) of the plant; such damage will allow entry of disease pathogens into the plant. Place the transplants immediately into their growing system.

Do not let them sit in direct light with their roots exposed as this will cause death of roots and some setback in the plants becoming re-established in the hydroponic system. If your hydroponic system is in a sunny patio location, transplant during the evening so that the plants can start to establish themselves overnight.

Tomato plants need to be spaced so as to allow sufficient light to enter the crop canopy (total upper growth of plants) and provide adequate air movement among the plants, achieved by proper spacing and training of the plants.

The correct spacing for tomatoes is 14 to 16 inches within the rows and 16 to 18 inches between the rows. They can be staggered in position from one another in adjacent rows to minimize leaf overlapping and to maximize their exposure to the light. They are planted in double rows with a 24 to 30-inch-wide aisle between these rows to allow access for plant care. This also helps in air movement and light penetration of the crop.

Training and Pruning

Tomatoes are trained vertically, retaining a single main stem. The most common method of supporting the stem is with plastic twine secured above to a support wire. While stakes could be used, they tend to fall over when the plants become heavy with fruit.

During transplanting, fasten the lower end of the twine to the tomato plant under a pair of large true leaves with a plastic clip. Plastic twine and stem clips are available from greenhouse and hydroponic suppliers. Leave several extra feet of length on the string above the support wire so that the plants may be lowered once they reach the height of the wire.

As the plant grows, wrap the twine in a clockwise direction around the stem. Make one rotation around the plant over the stem length of several leaves. If you always wrap the string in a clockwise direction, you will not mistakenly do the opposite in the future and inadvertently unwind the twine. This does happen with workers in commercial greenhouse operations in growing tomatoes and other vine crops.

Place the stem clamps every foot of plant

height. The twine is placed in the hinge of the clamp before closing it. Be sure to position the clamps immediately under the leaf petioles (leaf stem joining the main plant stem) as shown in Figure 33, not above the leaves as this will not hold the plant. Never place a clamp under a fruit cluster as it may puncture the fruit or cause the cluster to break.

Tomato plants are trained vertically by removing all of the side shoots or suckers that

Figure 33. Plastic stem support clamps placed immediately below leaf petioles.

grow between the main stem and each leaf petiole (Fig.34). They must be removed early when about two inches long. Otherwise, they drain the plant of nutrients and will make it quite vegetative, drawing nutrients into leaf and stem growth instead of fruit production.

(Tomatoes are highly acid and stain your hands when working with them. It is advisable to use disposable latex gloves when handling the plants.)

The suckers are removed by holding the stem just below the sucker with one hand while grasping the sucker with the other hand and giving it a fast jerk to the side. A clean break will occur as long as the suckers are relatively small.

Do not use a knife, pruning shears or scissors,

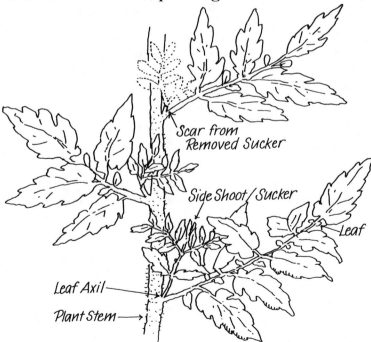

Figure 34. Tomato plant suckers (side shoots) at leaf axils.

as these tools will carry disease from one plant to another. Often the tops of the plants fork or split. Remove the less vigorous stem by breaking it off. If for some reason you badly tear the base of a sucker, use a knife to make a clean cut.

As the plants mature, generally by the time several clusters of fruit have been harvested, the lower leaves will become senescent (turn yellow). These leaves must be removed to allow better air circulation through the base of the plants and to prevent any disease infection.

Once the leaves have yellowed sufficiently, you will notice that a distinct line between the yellowing leaf petiole and the green stem has formed. This is a natural breaking point where, by hand, you can remove the yellow leaves. Again, hold the main stem with one hand and with the other grasp the leaf near its junction with the stem and give it a quick snap sideways. It should break clean. Once again, if you get a tear, make a clean cut with a knife or pruning shears.

Remove all of the leaves from the growing area as they may have fungal spores which could spread diseases to the rest of the plants.

Do not remove green leaves from the plants unless there is very poor air circulation along their bases. If you must remove some green leaves take off only one or two. Normally, remove leaves only below ripening fruit clusters.

It is now standard practice in the growing of tomatoes commercially in greenhouses to remove some of the fruit from the flower clusters when the fruit is less than one-fourth inch in diameter.

This is done to get more uniform size and ripening of the fruit. It also prevents breakage of the fruit cluster due to excessive weight of a large number of fruit on a cluster. Ideally, each cluster should have four to five fruit on it, ranging in size from the larger near the stem to the smallest at the extremity of the cluster (Figs.35,36,38). First, of course, remove mis-shapen, double or scarred fruit.

When the plants grow up to the support wire, which is located about seven feet above the base of the plants, they will have to be lowered. This is done by loosening the string at the wire and slowly lowering the plant by about a foot, while bending the lower part of the stem along the bed.

Figure 35. Pruned fruit cluster, for ideal fruit size
and shape development.

Figure 37. Removal of lower leaves and lowering of tomato plants.

Figure 36. Ripening clusters of fruit.

If this and the removal of lower leaves is done in the correct amount, about four to five feet of growth will remain on the main stem (Fig.37).

Pollination

Pollination is the process whereby the male pollen must be transferred to the female part of the flower (stigma of the style). Fertilization occurs and small bead-like fruit begin to develop within a week (fruit set).

Tomatoes growing outside are wind pollinated. Inside, pollination is done by hand.

Flowers are receptive (ready for pollination) when their petals begin to bend backwards (Fig.38). They remain in this state for several

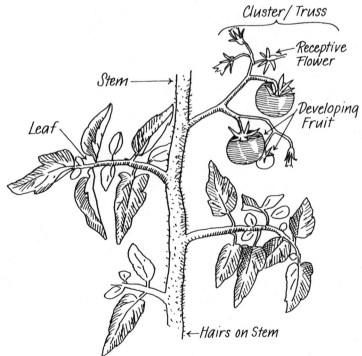

Figure 38. Receptive tomato flowers and "set" fruit on a flower cluster (truss).

days. For that reason, it is best to pollinate each day. You may pollinate by tapping the back of the flower or flower cluster to vibrate it. Battery-operated vibrators specifically for pollinating tomatoes are available from greenhouse suppliers. You will see the pollen flow like dust. The pollen will flow most readily during the late morning to early afternoon (11:00 AM to 3:00 PM). In the early morning, the relative humidity may be high and the pollen grains will stick together. Later in the day, if temperatures are very high, the pollen may dry.

Physiological and Nutritional Disorders

When environmental conditions are not right and/or nutrients are not in balance, plants will show specific symptoms of physiological disorders.

Poor light conditions, for example, reduce fruit set on tomatoes, cause the plant to be thin-stemmed and spindly, the leaves become light green in color. Poor light and high relative humidity resulting in poor pollination causes "catfacing" or distortion of the fruit (Fig.39).

The fruit may show uneven, "blotchy" ripening under low light, cool temperatures and high-nitrogen conditions. If the fruit is exposed to high temperatures and direct sunlight, "green shoulder" or "sunscald," coloration, showing green blotches and uneven ripening, may develop in certain varieties. This problem can easily occur if you prune too many lower leaves, especially above the ripening clusters of fruit.

Excessively high temperatures, above 90 degrees F., cause drying of pollen and reduce fruit set. High humidity can cause guttation (small droplets of water on the margins of the leaves) during the early morning.

"Blossom-end rot" (a sunken, leathery appearance on the blossom end of the fruit, often penetrating into the fruit as brown tissue) is caused by uneven watering, excessive water in the roots (puddling) resulting in root die-back or a calcium deficiency (Fig.40). High temperatures and infrequent watering cause fruit cracking (Fig.41).

Nutritional disorders occur in plants when one or more nutrients are at suboptimal levels. These

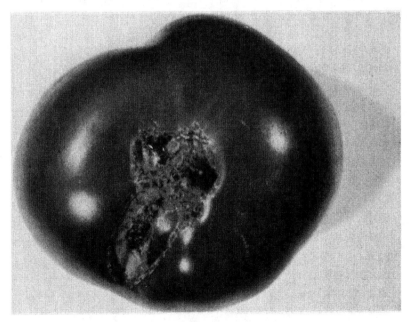

Figure 39. Catfaced tomato fruit.

Figure 40. Blossom-end-rot of tomato fruit.

disorders may be the result of a deficiency or excess of nutrients. In the early stages of a disorder, the plant will express specific symptoms of a particular disorder. However, as the disorder progresses, symptoms become more generalized and identification of the disorder is more difficult.

Therefore, it is important that you observe any changes in appearance of your plants. Watch the color of the leaves. If they turn yellow, or parts of them yellow (show chlorosis) and, later, brown spots develop or leaf margins brown (necrosis), something is probably wrong with the nutrient solution formulation. It is probably out of "balance." That is, the individual elements are not in the correct proportions or concentrations.

Recall what we said earlier about electrical

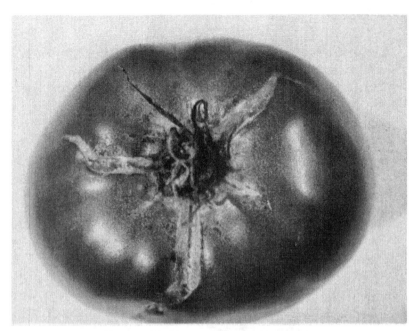

Figure 41. Tomato fruit cracking

conductivity (EC)—that it only measures total solutes, not levels of individual ions. So the nutrient solution may have the correct EC but still be umbalanced! If a nutritional disorder occurs, change the nutrient solution immediately. This will provide the correct balance of nutrients to the plant. If the problem persists, a change in formulation would be necessary. But this should be done only after obtaining a nutrient solution analysis from one of the commercial laboratories such as those listed in the Appendix.

This book does not explain the identification of nutritional disorders or how to interpret solution analyses and make adjustments in your formulations. Such procedures are explained in detail in *Hydroponic Food Production* (Woodbridge Press), see Appendix. There are a number of other publications listed in the Appendix that deal with nutritional disorders, pests and diseases of tomatoes including *Hydroponic Home Food Gardens* (Woodbridge Press).

Pests and Diseases

You may think that a hydroponic system should never have any pest or diseases. Not so. While hydroponics can greatly reduce the occurrence of diseases, particularly in the roots, if the system is sterile, it can also be conducive to the rapid spread of root diseases if they are introduced into the system.

Prevention and exclusion of diseases and pests is very important to minimize the occurrence of such problems. Keep the growing channels and tank covered to prevent algae growth; algae provides a home for some pests, such as fungus gnats. Keep the base of the plants dry, free of moisture—which is needed for the germination of fungal spores and hatching of insect eggs such as fungus gnats. By removing lower leaves that are becoming senecent (dying), allowing good air movement through the base of the plants, the relative humidity will be reduced in that area—especially at night and during early morning hours when moist conditions favor the germination of fungal spores.

As you gain experience with your plants, you will be able to identify many disease or pest problems.

Some of the more common ones are: tomato mosaic virus (TMV) which causes a distortion of the leaves (Fig.42) and reduction in overall plant vigor and productivity.

Any such plants should be immediately and carefully removed. Do not brush the infected plant onto another healthy one as that will transmit the disease. Also, wash your hands with soap and water before touching any other plants. Dispose of the infected plant in a garbage bag. The more recent greenhouse varieties of tomatoes have resistance to TMV.

Grey mold, or *Botrytis*, infects wounds or scars under high-humidity conditions. Most susceptible are scars remaining after leaf and sucker removal. A fluffy, grey growth forms at

Figure 42. TMV-infected tomato plant.

the infection site. Control it by reducing humidity, especially during early morning, or by using specific fungicides. Leaves may become infected with leaf mold—small grey spots develop on the undersides of leaves, especially overlapping lower leaves, or where sucking insects like white flies or aphids are present. Control leaf mold by reducing humidity and eliminating the insects.

Some of the more common insects on tomatoes are white flies, two-spotted spider mites, aphids, leafminers and fungus gnats. There are a number of books which outline the life cycles and identification of these insects. *Hydroponic Food Production* (Woodbridge Press) and others are listed in the Appendix.

Control measures are also presented in these books. Some insects can be controlled by safe insecticides such as "Safer's Soap" and pyrethrin. Alternatively, biological control agents or natural predators may be introduced on the tomato plants. These predators remain on the plants and eat the pests. The biological agents may be purchased from greenhouse and hydroponic suppliers. They are not a problem in your house or apartment as they will not move from the plants. The Appendix lists some excellent books on identification of pests and recommendations for their predators.

In summary, pests and diseases, if detected early, can generally be controlled without significant loss in the productivity of your garden.

Emphasis is on cleanliness, careful observation to detect the presence of the pests or diseases, correct identification and quick steps to control them. Between crops, sterilization of the beds, tank, equipment and irrigation lines with a 10 percent chlorine bleach solution will eliminate any carry-over of pest problems.

Never transplant plants grown in soil into your hydroponic system as this will certainly introduce pest problems and could even introduce weed seeds into the garden. Grow your seedlings in a sterile medium as discussed earlier. You will have clean plants in your hydroponic system.

Aerial parts of the tomato plants are susceptible to pests and diseases normally found in other cultural methods. The objective here is to identify them and control them early upon infestation. Maintain favorable environmental conditions to the plant; not those that favor disease or insects. For instance, high relative humidity favors fungal infection as mentioned earlier, and even some insects.

As you gain experience with your hydroponic gardening, you will seek more information on all aspects of it including some of the problems outlined in this guide. Whenever you encounter a problem, face the challenge of identifying it and realize the achievement in controlling it. That is what hydroponics offers: a sense of being able to control the growth and productivity of your tomatoes through optimization of nutrition and through the application of effective horticultural practices that foster environmental conditions favorable to plant growth. Regulation of pests through the use of biological agents, for example, helps you to experience the balance of a small ecological system.

When all of these factors are kept at levels

most beneficial to the tomatoes you will get maximum productivity as a reward for your good management of the hydroponic system (Fig.43). You will achieve the satisfaction of providing yourself and your family with high quality, clean, pesticide-free, highly nutritious tomatoes with that wonderful home-grown flavor!

Enjoyment in growing tomatoes hydroponically can stimulate an interest in horticulture generally, and inspire you to use hydroponics in growing other vegetables, even ornamentals, for the continuing pleasure of success.

Figure 43. Highly productive tomato plants with ripe fruit. The plants are growing in expanded clay medium.

Appendix

The Nutrient Solution

Fertilizer Salts

The most common fertilizer salts used to provide all essential elements of the nutrient solution are listed in Table 1.

Table 1: Fertilizer Salts and Their Essential Elements

Fertilizer	Chemical Formula	Elements Provided
A. Macroelements		
Ammonium Nitrate	NH_4NO_3	Nitrogen (N)
Calcium Nitrate	$Ca(NO_3)_2$	Calcium (Ca) Nitrogen (N)
Potassium Nitrate	KNO_3	Potassium (K) Nitrogen (N)
Monopotassium Phosphate	KH_2PO_4	Potassium (K) Phosphorus (P)
Potassium Sulfate	K_2SO_4	Potassium (K) Sulfur (S)
Magnesium Sulfate	$MgSO_4$	Magnesium (Mg) Sulfur (S)
Phosphoric Acid	H_3PO_4	Phosphorus (P)
B. Microelements		
Iron Chelate (10% iron) (Sequestrene 330)	FeEDTA	Iron (Fe)
Boric Acid	H_3BO_3	Boron (B)
Copper Sulfate	$CuSO_4$	Copper (Cu) Sulfur (S)
Manganese Sulfate	$MnSO_4$	Manganese (Mn) Sulfur (S)
Manganese Chloride	$MnCl_2$	Manganese (Mn) Chloride (Cl)
*Manganese Chelate Liquid (5% Mn)	MnEDTA	Manganese (Mn)
Zinc Sulfate	$ZnSO_4$	Zinc (Zn) Sulfur (S)
Zinc Chloride	$ZnCl_2$	Zinc (Zn) Chloride (Cl)

Table 1 (Continued):

*Zinc Chelate Powder (14% Zn) Liquid (9% Zn)	ZnEDTA	Zinc (Zn)
Sodium Molybdate	$Na_6Mo_7O_{24}$	Molybdenum (Mo) Sodium (Na)
Ammonium Molybdate	$(NH_4)_6Mo_7O_{24}$	Molybdenum (Mo) Nitrogen (N)

Alternatives, may be more difficult and expensive to purchase.

Nutrient Formulations:

Tomatoes require three levels of formulation during their various stages of growth. These levels are expressed as A, B and C, increasing in strength with plant maturity. Table 2 presents the three formulations for tomatoes.

Note that the formulations assume that the raw water has no nutrients in it. As explained earlier, some adjustments will have to be made if your water contains any of the elements. For example, if your raw water contains 30 ppm of magnesium (Mg) and 50 ppm of calcium (Ca), subtract these amounts from the formulation and add the remainder. If the concentration of any element in the raw water exceeds that required in the nutrient formulation, do not add any of that element to the nutrient solution.

Assume that your nutrient tank has a volume of 100 liters (26 US gal) for convenience to demonstrate the calculations involved:

Formulation A of Table 2 calls for 20 ppm of Mg and 100 ppm of Ca. Clearly, the residual level of 30 ppm of Mg in the raw water exceeds that required by the nutrient formulation, so do not add any magnesium sulfate to the solution. You must still add 100 – 50 ppm = 50 ppm of calcium, in the form of calcium nitrate.

In formulation B you need to add 33 – 30 = 3 ppm of Mg and 150 – 50 = 100 ppm of Ca. Formulation C lacks 45 – 30 = 15 ppm of Mg and 200 – 50 = 150 ppm of Ca.

Table 2: Tomato Nutrient Formulations A, B & C
for Three Stages of Plant Maturity, with Fertilizer Source

Formulation Level	Crop Maturity	Fertilizer	Elemental Formulation (ppm)
		A. Macroelements	
A (Stage 1)	Seedlings from first true-leaf stage (10-14 days old) until plants are 14"-16" tall.	Calcium Nitrate Potassium Nitrate Monopotassium Phosphate Potassium Sulfate Magnesium Sulfate	N – 100 P – 40 K – 200 Ca – 100 Mg – 20 S – 53
B (Stage 2)	14"-16" tall until 24" tall when fruit set on first cluster is 1/4" to 1/2" in diameter.	Calcium Nitrate Potassium Nitrate Monopotassium Phosphate Potassium Sulfate Magnesium Sulfate	N – 130 P – 55 K – 300 Ca – 150 Mg – 33 S – 109
C (Stage 3)	After first fruit set to plant maturity.	Calcium Nitrate Potassium Nitrate Monopotassium Phosphate Potassium Sulfate Magnesium Sulfate	N – 180 P – 65 K – 400 Ca – 400 Mg – 45 S – 144
		B. Microelements	
A, B & C	Use the same formulation for all three levels of plant maturity.	Boric Acid Manganese Sulfate Copper Sulfate Zinc Sulfate Sodium Molybdate Iron Chelate (10% iron)	B –0.3 Mn –0.8 Cu –0.07 Zn –0.10 Mo –0.03 Fe –3.0

Note: Only the *macroelements* have three levels of formulations, the *microelements* are the same for all three levels.

For Formulation C the calculations are as follows, using Tables 2 and 3. Mg: 15 ppm additional required. Amount needed (ppm)/total (ppm) x total weight of fertilizer: 15 ppm/45 ppm x 45 g = 15 g of magnesium sulfate. Ca: 150 ppm is lacking. 150 ppm/200 ppm x 92 g = 69 g of calcium nitrate.

Now, since we are using less calcium nitrate, our

nitrogen level will also be reduced: weight of adjusted nutrient compound (g)/weight of full formulation nutrient compound (g) x total concentration of nitrogen (ppm) in normal formulation = adjusted level of nitrogen (ppm): 69 g/92 g x 140 ppm = 105 ppm of nitrogen (N) from calcium nitrate.

The difference of 140 ppm – 105 ppm = 35 ppm of N must be added through another fertilizer. You could adjust upward the amount of potassium nitrate, but that will add potassium, which will have to be adjusted for by reducing the potassium sulfate proportionately. An easier method is to obtain the 35 ppm deficit in nitrogen by using ammonium nitrate. From calculations not shown, involving a conversion factor, the amount of ammonium nitrate required to supply 35 ppm of N in a 100 liter tank is 10 grams.

Table 2 outlines the various stages of plant growth for each level of formulation, the fertilizer salt as the source of the various elements and elements provided expressed in parts per million.

The elements are divided into two groups, *macroelements*, those required in relatively large amounts and *microelements*, those needed in small amounts.

Table 3 gives the amounts of fertilizer salts for each formulation required for several nutrient tank solution volumes. From this table you may extrapolate the amounts you would need for other tank volumes. For example, in the seedling stage of plant growth the amount of calcium nitrate needed in a tank of 100 liters (26 U.S. gallons) is 46 grams. If your tank had a volume of 25 liters (6.5 U.S. gallons), you would want: 25/100 x 46 grams = 11.5 grams of calcium nitrate.

Table 3: Fertilizer Weights (grams) to Supply the Nutrients of Three Formulation Levels for Tomatoes with Three Nutrient Tank Volumes.

Formulation Level	Fertilizer	g/100 l (26 US gal)	g/50 l (13 US gal)	g/38 l (10 US gal)	Elements Provided (ppm)
	A. Macroelements				
A	Calcium Nitrate	46	23	17.7	Ca –100 N– 70
	Potassium Nitrate	23	11.5	8.85	K– 84 N– 30
	Monopotassium Phosphate	18	9	6.92	P– 40 K– 50
	Potassium Sulfate	16	8	6.15	K– 66 S– 27
	Magnesium Sulfate	20	10	7.70	Mg– 20 S– 26
B	Calcium Nitrate	69	34.5	26.5	Ca –150 N–105
	Potassium Nitrate	19	9.5	7.31	K– 70 N– 25
	Monopotassium Phosphate	25	12.5	9.62	P– 55 K– 69
	Potassium Sulfate	39	19.5	15.0	K–161 S– 66
	Magnesium Sulfate	33	16.5	12.7	Mg– 33 S– 33
C	Calcium Nitrate	92	46	35.4	Ca –200 N–140
	Potassium Nitrate	31	15.5	11.9	K–112 N– 40
	Monopotassium Phosphate	29	14.5	11.2	P– 65 K– 81
	Potassium Sulfate	50	25	19.2	K–207 S– 85
	Magnesium Sulfate	45	22.5	17.3	Mg– 45 S– 58
	B. Microelements				
A, B & C	Boric Acid	0.17	0.085	0.065	B–0.3
	Manganese Sulfate	0.32	0.16	0.123	Mn–0.8
	Copper Sulfate	0.028	0.014	0.011	Cu–0.07
	Zinc Sulfate	0.045	0.0225	0.017	Zn–0.10
	Sodium Molybdate	0.013	0.0065	0.005	Mo–0.03
	Iron Chelate (10% iron)	3.0	1.5	1.15	Fe–3.0

See Note on Next Page

Note to Table 3, Preceding Page: Only the macroelements have three levels of formulations, the microelements are the same for all three levels. These quantities of the various fertilizers supplying the three formulations assume that the raw water is free of any residual elements (same as distilled water). If your raw water contains some elements, you must have it analyzed and make adjustments to your formulation as described in the section "Nutrient Formulations."

Nutrient Solution Preparation

To avoid any chemical reactions which may result in precipitation (the coming out of solution in a solid, insoluble state), the fertilizer salts should be added in a specific sequence as outlined below:

1. Fill the nutrient tank one-third full with water.
2. Dissolve each fertilizer salt alone in a separate pail of water before adding it to the nutrient tank. Be careful not to use so much water that when you add it to the tank it will overflow.
3. Dissolve the fertilizers in the following sequence: ammonium nitrate, potassium nitrate, calcium nitrate and monopotassium phosphate.
4. Now fill the tank to three-quarters full, if it is not already at that level.
5. Dissolve potassium sulfate in a pail of water and add to the tank.
6. Check the pH of the solution and add acid or base to bring the pH to between 6.3 and 6.5. *(Note Pages 86-88.)*
7. Dissolve the micronutrients, including iron chelate, together in a bucket of water and then add them to the tank.
8. Top up the tank with water to the full level and check the pH again. Adjust it if necessary. *(Note Pages 86-88.)*

Nutrient Stock Solutions

Your home hydroponic system often has a nutrient tank with a volume of less than 50 gallons. As a result, you will require very small amounts of these fertilizer salts, especially the micronutrients. Some of the weights of these salts will be less than 1/100 of a gram. To achieve this accurately, you would need a very sensitive analytical balance which could cost over $1000.

To avoid the need for such equipment and to reduce error in weighing the fertilizers, make up a stock solution that is 100 to 200 times the strength of the final nutrient solution. This stock solution will contain 100 to 200 times the weight of each fertilizer that is needed in the regular nutrient solution.

Table 4 gives the weights of the various fertilizers needed to make up a 100 times and 200 times strength stock solution for the three tomato formulations. Note that two separate stock solutions must be made because the sulfates cannot be mixed with calcium nitrate since they would cause its precipitation. A third stock solution of 300 times normal concentration is used for the micronutrients; their weight is very small.

To make the nutrient solution from the stock solutions, dilute them back proportionally to the specified normal levels for the end nutrient solution. For example, if your nutrient tank has a volume of 100 liters (26 U.S. gallons) and stock solutions A and B are 200 times and the micronutrients are 300 times normal levels, prepare the nutrient solution as follows:

1. Fill the nutrient tank three-quarters full with raw water (75 liters).

2. With a graduated cylinder (available from scientific supply companies), measure: final nutrient tank volume divided by the stock solution strength (100/200) = 0.50 liters (500 milliliters) of solution A and add this amount to the nutrient tank.

3. Add more water to the tank to about 90 liters, then add 500 ml of stock solution B.

4. Check the pH and adjust it within the range of 6.3 - 6.5, using an acid or base.

5. Add the correct amount of micronutrient stock solution.

It was 300 times normal level; so, measure: 100/300 = 0.33 liters or 333 milliliters (ml), and add it to the tank.

6. Finally, top up the tank to 100 liters with raw water, then check and adjust the pH again, if necessary.

Table 4: Nutrient Stock Solutions for Three Levels of Tomato Formulations Using Various Solution Concentrations

Formulation Level	Fertilizer	Wt./100 l (26 US gal)	Wt./50 l (13 US gal)	Wt./38 l (10 US gal)	Wt./19 l (5 US gal)
	A. Macroelements (100X)				
A	*Part A:*				
	Calcium Nitrate	4.6 kg	2.3 kg	1.77 kg	885 g
	Part B:				
	Potassium Nitrate	2.3 kg	1.15 kg	885 g	442 g
	Monopotassium Phosphate	1.8 kg	0.9 kg	692 g	346 g
	Potassium Sulfate	1.6 kg	0.8 kg	615 g	308 g
	Magnesium Sulfate	2.0 kg	1.0 kg	769 g	348 g
B	*Part A:*				
	Calcium Nitrate	6.9 kg	3.45 kg	2.65 kg	1.325 kg
	Part B:				
	Potassium Nitrate	1.9 kg	0.95 kg	731 g	366 g
	Monopotassium Phosphate	2.5 kg	1.25 kg	962 g	481 g
	Potassium Sulfate	3.9 kg	1.95 kg	1.50 kg	750 g
	Magnesium Sulfate	3.3 kg	1.65 kg	1.27 kg	635 g
C	*Part A:*				
	Calcium Nitrate	9.2 kg	4.6 kg	3.54 kg	1.77 kg
	Part B:				
	Potassium Nitrate	3.1 kg	1.55 kg	1.19 kg	595 g
	Monopotassium Phosphate	2.9 kg	1.45 kg	1.12 kg	560 g
	Potassium Sulfate	5.0 kg	2.6 kg	1.92 kg	960 g
	Magnesium Sulfate	4.5 kg	2.25 kg	1.73 kg	865 g
	A. Macroelements (200X)				
A	*Part A:*				
	Calcium Nitrate	9.2 kg	4.6 kg	3.54 kg	1.77 kg
	Part B:				
	Potassium Nitrate	4.6 kg	2.3 kg	1.77 kg	885 g
	Monopotassium Phosphate	3.6 kg	1.8 kg	1.38 kg	690 g
	Potassium Sulfate	3.2 kg	1.6 kg	1.23 kg	615 g
	Magnesium Sulfate	4.0 kg	2.0 kg	1.54 kg	770 g
B	*Part A:*				
	Calcium Nitrate	13.8 kg	6.9 kg	5.30 kg	2.65 kg
	Part B:				
	Potassium Nitrate	3.8 kg	1.9 kg	1.46 kg	730 g
	Monopotassium Phospate	5.0 kg	2.5 kg	1.92 kg	960 g
	Potassium Sulfate	7.8 kg	3.9 kg	3.00 kg	1.50 kg
	Magnesium Sulfate	6.6 kg	3.3 kg	2.54 kg	1.27 kg

Table 4 (Continued):

C	Part A:				
	Calcium Nitrate	18.4 kg	9.2 kg	7.08 kg	3.54 kg
	Part B:				
	Potassium Nitrate	6.2 kg	3.1 kg	2.38 kg	1.19 kg
	Monopotassium Phosphate	5.8 kg	2.9 kg	2.23 kg	1.12 kg
	Potassium Sulfate	10.0 kg	5.0 kg	3.85 kg	1.92 kg
	Manganese Sulfate	9.0 kg	4.5 kg	3.46 kg	1.73 kg

B. Microelements (300X)

A, B & C	Boric Acid	51.0 g	25.5 g	19.6 g	9.80 g
	Manganese Sulfate	96.0 g	48.0 g	36.9 g	18.45 g
	Copper Sulfate	8.4 g	4.2 g	3.23 g	1.62 g
	Zinc Sulfate	13.5 g	6.75 g	5.19 g	2.60 g
	Sodium Molybdate	3.9 g	1.95 g	1.50 g	0.75 g
	Iron Chelate (10% Iron)	900.0 g	450.0 g	346.0 g	173.0 g

Note: Iron chelate is added last after adjusting the pH to 5.5 – 6.0.

If the tank volume is not exactly as one of those listed in Table 3 or 4, you can extrapolate as follows: From Table 3, for example, determining the nutrient weights under Formulation A of the macroelements: Use the fraction of your tank volume over the one shown in the table and multiply it times the weight for each fertilizer of the formulation.

Thus, calcium nitrate for a 13 U.S. gallon tank will weigh 23 grams. Therefore, a 7.5 U.S. gallon tank will require: 7.5/13 x 23 = 13.3 grams of calcium nitrate. Do this for each fertiizer to get the amounts for the entire formulation.

Remember to always check the pH and adjust it before adding the micronutrient stock solution.

Keep all stock solutions in a safe place away from children and animals, and store it in an opaque container so that light will be excluded in order to prevent the growth of algae.

Suppliers

Note: This is not a complete list of suppliers of these products.

Biological Control Agents:

Applied Bio-Nomics Ltd.
P.O. Box 2637
Sidney, B.C.
Canada V8L 4C1

Beneficial Bugs
P. O. Box 1627
Apopka, FL 32703-1627

Foothill Hydroponics
10705 Burbank Blvd.
N. Hollywood, CA 91601

Hydro-Gardens Inc.
P. O. Box 9707
Colorado Springs, CO 80932

Koppert (UK) Ltd.
Biological Control
P.O. Box 43
Tunbridge Wells
Kent TN2 5BX
United Kingdom

Nature's Control
P.O. Box 35
Medford, OR 97501

Organic Pest Management
P.O. Box 55267
Seattle, WA 98155

Rincon-Vitova Insectories
P.O. Box 95
Oak View, CA 93022

Hydroponic Consultants & Technology

Agro-Dynamics Inc.
12 Elkins Rd.,
East Brunswick, NJ 08816

Howard M. Resh, Ph.D.
International Aquaponics
P.O. Box 1161
Kelowna, BC
Canada V1Y 7P8

Hydroponic and Soilless Culture Society

Hydroponic Society of America
P.O. Box 6067
Concord, CA 94524

Hydroponic Equipment & Supplies

Agro-Dynamics, Inc.
12 Elkins Rd.
East Brunswick, NJ 08816

American Hort. Supply
25603 W. Ave. Stanford
Valencia, CA 91355

American Horticultural Supply
4045 Via Pescador
Camarillo, CA 93012

American Produce
P.O. Box 123
Arcola, VA 22010

Applied Hydroponics, Inc.
3135 Kerner Blvd
San Rafael, CA 94901

Applied Hydroponics, Canada
2215 Walkley
Montreal, Quebec
Canada H4B 2J9

Applied Hydroponics, Phil.
208 Rt. 13
Bristol, PA 19007

Aqua-Culture, Inc.
P.O. Box 26467
Tempe, AZ 85282

Aqua-Ponics, Inc.
P.O. Box 41136
Los Angeles, CA 90041-1224

Berkeley Indoor Garden Center
844 University Ave.
Berkeley, CA 94710

Brazos Indoor Garden Supply
P.O. Box 192
Houston, TX 77274

Raymond Bridwell
P.O. Box 192
Perris, CA 92370

Brisbon Enterprises
209 Riley Drive
Pacheco, CA 94553

Central Washington
  Halide & Hydroponics
1001 Fruitvale Blvd.
Yakima, WA 98902

Canadian Hydroponic
Information Service
1648 Ave. Rd.
Toronto, Ont., Canada M5M 3Y1

Canadian Hydrogardens Ltd.
411 Brook Rd. West
Ancaster, Ontario
Canada L9G 3L1

Canadian Hydroponics, Ltd.
8318 120th St.
Surrey, BC
Canada V3W 3N4

Canadian Hydrophyte Systems
2402 Edith Ave.
Burlington, Ont.
Canada L7R 1N6

CESCO
Controlled Environmental Syst.
18677 Westview Dr., #200
Santa Clara, CA 95070

Dr. Chatelier's Plant Food
400 Douglas Road East
Oldsmar, FL 34677

Clover Greenhouses
P.O. Box 789
Smyrna, TN 37167

CropKing, Inc.
P.O. Box 310
Medina, OH 44258

Diamond Lights & Hydroponics
713 Mission Ave. E.
San Rafael, CA 94901

Disc. Halide & Hydroponics
14109 Sprague, Suite 5
Spokane, WA 99216

Dyna Grow Corp.
372 E. Bell Marin Keyes
Novato, CA 94947

East Coast Hydroponics, Inc.
432 Castleton Ave.
Staten Island, N.Y. 10301

Eco Enterprises
2821 N.E. 55th St.
Seattle, WA 98105

Edison Lighting & Wholesale.
1221 S.E. Water Ave.
Portland, OR 97214

Electro-Tec of Albuquerque
7504 Dellwood Rd., N.E.
Albuquerque, NM 87110

Ellwood Greenhouses
5074 E. Cooper St.
Tucson, AZ 85711

Engineered Systems and Designs
3 South Tatnall St.
Wilmington, DE 19801

ENP
200 North Main St.
Mendota, IL 60507

Florida Home Grown Hydroponics
124 Robin Rd., Suite 1300
Altamonte Springs, FL 32701

Florida Indoor Growers Supply
7555 S.W. E. Cooper St.
South Miami, FL 33143-4616

Foothill Hydroponics
10705 Burbank Blvd.
North Hollywood, CA 9160

Future Garden Supply, Inc.
12507 Pacific Ave.
Tacoma, WA 98444

Genova Products
7034 East Court St.
Davison, MI 48423

W.R. Grace & Co.
P.O. Box 798
Iron Run Industrial Park
Fogelsville, PA 18051

Grow-Max Systems, Inc.
10070 McIntosh Rd.
Dover, FL 335 27

Hahn's Lighting
945 South 3rd St.
San Jose, CA 95112

Hamilton Technology Corp.
14902 South Figueroa St.
Gardena, CA 90248

Hollister's Hydroponics
P.O. Box 16601
Irvine, CA 92713

Homegrown, Inc.
12605 Pacific Ave.
Tacoma, WA 98444

Hydro-Gardens, Inc.
P.O. Box 9707
Colorado Springs, CO 80932

Hydrolights
347 Nord Ave., #1
Chico, CA 95926

Hydroponic Sales & Services
159 South Rd.
Ridleyton 5008
Australia

Hydro-Tech
3929 Aurora Ave., N.
Seattle, WA 98103

Interior Water Garden
615 Long Beach Blvd.
Surf City, NJ 08008

Full Moon Farm Products
P.O. Box 2046
1217 S.W. 2nd
Corvalis, OR 97339

General Hydroponics
124 San Luis Way
Novato, CA 94945

Geotechnology
1035 17th Ave.
Santa Cruz, CA 95062

The Greenhouse & Hydro Store
63 Clarke Sideroad # 17
London, Ont.
Canada N5W 5W7

Grotek
1/403 Princes Hwy.
Noble Park, Victoria 3174
Australia

Halide of Oregon
9220 S.E. Stark
Portland, OR 97216

Higher Yield
3811 N.E. 292 Ave.
Camas, WA 98607

Home Harvest Garden Supply
1324 Jefferson Davis Hwy.
Woodbridge, VA 22191

A.H. Hummert Seed Co.
2746 Chouteau Ave.
St. Louis, MO 63103

Hydroponic Growing Systems
32 Richardson Rd.
Ashby, MA 01431

Hydro-Pioneer, Inc.
8672 W. Hindsdale Pl.
Littleton, CO 80123

Hydroponic Technologies
1313 Ritchie Ct., Suite 2401
Chicago, IL 60610

The Indoor Gardener
1311 South Pacific Hwy.
Talent, OR 97540

Kwik-Grow Systems
16515 S.W. Rigert Terrace
Beaverton, OR 97007

Leisure Garden
P.O. Box 5038
Winston-Salem, NC 27113

Living Green, Inc.
4091 E. La Palma Ave., Suite E
Anaheim, CA 92807

MAH
151 Alkier St.
Brentwood, NY 11717

Master Blend
P.O. Box 0329
Gobbles, MI 49055

Mellinger's
2310 W. South Range Rd.
North Lima, OH 44452

Midwest Perlite, Inc.
Rt. 9, 4280 W. Parkway Blvd.
Appleton, WI 54915

Northwest Seed & Pet
East 2422 Sprague
Spokane, WA 99202

New England Hydroponics
90 Outlook Dr., Apt. 34
Worcester, MA 01602

New Zealand Hydroponics, Ltd.
P.O. Box 949
Tauranga, New Zealand

Northern Hydroponics, Ltd.
12723 Fort Rd.
Edmonton, Alta., Canada T5A 1A7

North Star Lights, Hydroponics
5200 Euclid Ave
Cleveland, OH 44103

Plantermation
34 Lakeshore
Irvine, CA 92714

Rain Or Shine Garden Supplies
13126 N.E. Airport Way
Portland, OR 97230

Reed Company
16714 Meridian St. #239
Puyallup, WA 98373

Light Manufacturing Co.
1634 S.E. Brooklyn
Portland, OR 97202

Lutz Hydroponics
18014 Crooked Lane
Lutz, FL 33549

McCalif
2215 Ringwood Ave.
San Jose, CA 95131

J.M. McConkey & Co.
P.O. Box 309
Sumner, WA 98390

Midwest Growers Supply
2613 Kaneville Court
Geneva, IL 60134

Natural Liquid Fertilizer Corp.
3724 W. 38th St.
Chicago, IL 60632

New Earth Indoor/Outdoor
   Garden Center
4422 East Hwy. 44
Shepherdsville, KY 40165

N.W. Agricultural Supply
P.O. Box 975
Hermiston, OR 97838

New Garden Supply
Highway 10 West
Livingston, MT 59047

Northern Lights & Hydroponics
4 Mahoning Ave.
New Castle, PA

Paradise Lighting, Hydroponics
306 West Harris
Eureca, CA 95501

P.L. Light Systems, Canada
183 S. Service Rd., Unit 2
Grimsby, Ontario
Canada L3M 4G3

Rambridge Structure & Design
1316 Centre St., N.E.
Calgary, Alta., Canada T2E 2A7

Rehau Plastics, Inc.
P.O. Box 1706
Leesburg, VA 22075

Reindeer's Place
6700 S. Island Hwy.,P.O. Box 11
Bowser, B.C. Canada V0R 1G0

R & S Enterprises
ics
P.O. Box 7197
Huntsville, TX 77342

San Diego Hydroponics
213 Church Ave.
Chula Vista, CA 92010

Smithers-Oasis
P.O. Box 1204
Carpinteria, CA 93013

Sophisticated Systems
3785 U.S. Alt. 19 Nort
Palm Harbor, FL 33563

Suncor Systems, Inc.
P.O. Box 11116
Portland, OR 97211

Trojan Technologies, Inc.
845 Consortium Ct.
London, Ont.,Canada N6E 2S8

Urban Tek Growers Supply
2911 W. Wilshire Blvd.
Oklahoma City, OK 73116

Jack Van Klaveren Ltd. (JVK)
P.O. Box 910
St. Catharines, Ontario
Canada L2R 6Z4

Virginia Hydroponics
16 Bannister Dr.
Hampton, VA 23666

Western Water Farms
2803 Shaughnessy St.
Port Coquitlam, B.C.
Canada V3C 3H1

Worm's Way Florida
4402 N. 56th St.
Tampa, FL 33610

Worm's Way Mass.
1200 Milbury St., Suite 8-G
Worcester, MA 01607

Reko bv
P.O. Box 191
6190 AD Beek (L)
The Netherlands

Rocky Mt. Lighting, Hydroponics
7100 N. Broadway, Suite 55
Denver, CO 80221

Sharp & Son
19219 62nd Ave. S. Bldg. C.
Kent, WA 98032

Solar Hydroponic Supply
4779 Kingsway
Burnaby, B.C., Canada

Southern Lights & Hydroponics
5634 Buford Hwy.
Norcross, GA 30071

Superior Growers Supply
4870 Dawn Ave.
East Lansing, MI 48823

Troy Hygro-Systems, Inc.
4096 Hwy. ES
East Troy, WI 53120

Utelite Corp.
9160 S. 300 W. #22
Sandy, UT 84070

Vegetron, Hawaii
4475 Sierra Dr.
Honolulu, HI 96816

Westbrook Greenhouse Systems
270 Hunter Rd.
Grimsby, Ont.,Canada L3M 5G1

Westmark Company
3529 Touriga Dr.
Pleasanton, CA 94566

Worm's Way Indiana
3151 S. Hwy. 446
Bloomington, IN 47401

Worm's Way Missouri
12156 Lackland Rd.
St. Louis, MO 63146

Seeds

De Ruiter Seeds, Inc.
P.O. Box 20228
Columbus, OH 43220

Goldsmith Seeds, Inc.
P.O. Box 1349
Gilroy, CA 95021

Harris Seeds
P.O. Box 22960
Rochester, NY 14692-2960

Park Seed Co., Wholesale Div.
FS 231 Cokesbury Rd.
Greenwood, SC 29647-0001

Pan American Seed Co.
P.O. Box 4206
Saticoy, CA 93004

Stokes Seeds, Inc.
P.O. Box 548
Buffalo, NY 14240

Vaughan's Seed Co.
5300 Katrine Ave.
Downers Grove, IL 60515

G.S. Grimes/German Seeds,
201 W. Main St.
Smethport, PA 16749

A.H. Hummert Seed Co.
2746 Chouteau Ave.
St. Louis, MO 63103

Kube-Pak Corp.
R.D. #3, Box 2553, Rt. 526,
Allentown, NJ 08501

Penn State Seed Co.
Route 309, P.O. Box 390
Dallas, PA 18612

Sluis & Groot
4600 S. Ulster St., St. 700
Denver, CO 80237

Stokes Seeds Ltd.
39 James St., Box 10
St. Catharines, Ont.,
Canada L2R 6R6

Zwaan Seeds, Inc.
P.O. Box 397
Woodstown, NJ 08098

Scientific Supply

Cole-Parmer Instrument Co.
7425 North Oak Park
Chicago, IL 60648

GIBCO Laboratories
3175 Staley Rd.
Grand Island, NY 14072

GIBCO Canada
2260A Industrial Ave.
Burlington, Ontario
Canada L7P 1A1

Myron L Co.
6231 C Yarrow Dr.
Carlsbad, CA 92009

Fisher Scientific
50 Fadem Rd.
Springfield, NJ 07081

GIBCO Laboratories
519 Aldo Ave.
Santa Clara, CA 95050

Hach Company
1313 Border St., Unit 34
Winnipeg, Manitoba
Canada R3H 0X4

Markson
7815 S. 46th St.
Phoenix, AZ 85044-4399

Soil and Plant Tissue Testing Laboratories

Dept. of Land Resource Science
University of Guelph
Guelph, Ont., Canada

Griffin Laboratories
1875 Spall Rd.
Kelowna, B.C., Canada V1Y 4R2

Micro Essential Laboratory Inc.
4224 Ave. H
Brooklyn, NY 11210

Micro-Macro International,

183 Paradise Blvd., Suite 108
Athens, GA 30607

Ohio State University
Ohio Agricultural Research and
Development Center
Research-Extension Analytical
Laboratory
Wooster, OH 44691

Soil & Plant Laboratory, Inc.
P.O. Box 11744
Santa Ana, CA 92711
P.O. Box 153
Santa Clara, CA 95052
P.O. Box 1648
Bellevue, WA 98009

Soil Testing Laboratory
Purdue University
Agronomy Dept.
Lafayette, IN 47907

Soil Testing Laboratory
Texas A & M University
College Station, TX 77843

Trade Magazines and Periodicals

American Vegetable Grower
Meister Publ. Co.
Willoughby, OH 44094

Greenhouse Grower
Meister Publ. Co.
Willoughby, OH 44094

Greenhouse Canada
Cash Crop Farming Publ. Ltd.
Delhi, Ont., Canada

GreenhouseManager
Box 1868
Fort Worth, TX 76101

Growing Edge Magazine
215 S.W. 2nd St.
Corvallis, OR 97333

Grower Talks
Geo. J. Ball, Inc.
West Chicago, IL 60185

The 21st Century Gardener
Growers Press, Inc.
P.O. Box 189
Princeton, B.C.
Canada V0X 1W0

The Herb Market Report
1305 Vista Dr.
Grants Pass, OR 97527

Houseplant Forum Magazine
1449 Ave. William
Sillery, Que., Canada G1S 4G5

Bibliography

Horticulture

Golden Nature Guide. *Insect Pests.* New York: Golden Press.

Golden Science Guide. *Botany.* New York: Golden Press.

Jarvis, W.R. and C.D. McKenn. 1984. *Tomato diseases.* Canada Dept. of Agric. Publ. 1479/E. Agric. Canada, Ottawa, Ontario.

Malais, M. and W.J. Ravensberg. 1992. *Knowing and recognizing-the biology of glasshouse pests and their natural enemies.* Koppert B.V.,Berkel en Rodenrijs, the Netherlands.

Nelson, Paul V. 1985. *Greenhouse operation and management.* 3rd ed. Reston, Virginia: Reston Publ. Co.

Roorda van Eysinga, J.P.N.L. and K.W. Smilde. 1981. *Nutritional disorders in glasshouse tomatoes, cucumbers, and lettuce.* Wageningen: Center for Agric. Publ. and Documentation.

Steiner, M.Y. and D.P. Elliott. 1983. *Biological pest management for interior plantscapes.* Min. of Agric. and Food, Victoria, B.C., Canada.

Wittwer, S.H. and S. Honma. 1979. *Greenhouse tomatoes, lettuce, and cucumbers.* East Lansing, Michigan: Michigan State Univ. Press.

Hydroponics

Bentley, M. 1974. *Hydroponics plus.* Sioux Falls, South Dakota: O'Connor Printers.

Bridwell, R. 1990. *Hydroponic gardening,* rev. ed. Santa Barbara, Calif: Woodbridge Press.

Cooper, A.J. 1979. *The ABC of NFT.* London: Grower Books.

Dalton, L. and R. Smith. 1984. *Hydroponic gardening.* Auckland: Cobb/Horwood Publ.

Douglas, J.S. 1973. *Beginner's guide to hydroponics.* London: Pelham Books.

Drakes, G.D., Sims, T.V., Richardson, S.J. and D.M. Derbyshire 1984. *Tomato production 6–Hydroponic growing systems.* Min. of Agric., Fisheries and Food. ADAS Booklet 2249. London: MAFF Publ.

Harris, D. 1974. *Hydroponics: The gardening without soil.* 4th ed. Capetown: Purnell.

Hudson, J. 1975. *Hydroponic greenhouse gardening.* Garden Grove, Calif.: National Graphics, Inc.

Johnson, H. Jr., G.J. Hochmuth, and D.N. Maynard. 1985. *Soilless culture of greenhouse vegetables.* Cooperative Ext. Service, Univ. of Florida, IFAS Bulletin 218. Gainsville, Florida: C.M. Hinton, Publ. Distrib. Center.

Jones, L. 1977. *Home hydroponics...and how to do it!* Pasadena, Calif.: Ward Ritchie Press.

Kenyon, S. 1979. *Hydroponics for the home gardener.* Toronto, Ontario: Van Nostrand Reinhold Ltd.

Maas, E.F. and R.M. Adamson. 1971. *Soilless culture of commercial greenhouse tomatoes.* Publ. 1460. Information Div., Canada Dept. of Agric., Ottawa, Ontario, Canada.

Resh, H.M. 1989. *Hydroponic food production.*, 4th ed. Santa Barbara, Calif.: Woodbridge Press.

Resh, H.M. 1990. *Hydroponic home food gardens.* Santa Barbara, Calif.: Woodbridge Press.

Schippers, P.A. 1977. *Construction and operation of the nutrient flow technique for growing plants.* Veg. Crops Mimeo 187. Cornell Univ., Ithaca, N.Y.

Schippers, P.A. 1979. *The nutrient flow technique.* Veg. Crops Mimeo 212. Cornell Univ., Ithaca, N.Y.

Smith, D.L. 1987. *Rockwool in horticulture.* London: Grower Books.

Sundstrom, A.C. 1989. *Simple Hydroponics—for Australian and New Zealand gardeners.* 3rd ed. South Yarra, Victoria, Australia: Viking O'Neil.

Sutherland, S.K. 1986. *Hydroponics for everyone.* South Yarra, Victoria, Australia: Hyland House.